Evolution of Social Ties Around New Food Practices

SCIENCES

Agronomy and Food Science, Field Directors –
Jack Legrand and Gilles Trystram

Food Chain Management, Subject Head – Jean-Marc Ferrandi

Evolution of Social Ties Around New Food Practices

Coordinated by
Gilles Séré de Lanauze

iSTE

WILEY

First published 2021 in Great Britain and the United States by ISTE Ltd and John Wiley & Sons, Inc.

ISTE Ltd
27-37 St George's Road
London SW19 4EU
UK

www.iste.co.uk

John Wiley & Sons, Inc.
111 River Street
Hoboken, NJ 07030
USA

www.wiley.com

Library of Congress Control Number: 2021943145

British Library Cataloguing-in-Publication Data
A CIP record for this book is available from the British Library
ISBN 978-1-78945-044-6

ERC code:
LS9 Applied Life Sciences, Biotechnology, and Molecular and Biosystems Engineering
LS9_5 Food sciences (including food technology, food safety, nutrition)

MIX
Paper from responsible sources
FSC® C013604

Contents

Foreword

Thibaut NGUYEN
Trends & Prospective Department, Ipsos, Paris, France

As becomes clearer every day, we are living in a period of unprecedented changes and challenges. For the last 50 years, we have been witnessing the progressive erosion of all the major collective reference points, together with an extraordinary opening of the field of scientific and technological possibilities.

This double movement of collapse and innovation is generating a groundswell that is disrupting our lifestyles, redefining daily life in its most essential acts: working, moving, communicating, eating.

The work carried out by Gilles Séré de Lanauze and the group he coordinates clearly and rigorously reveals the extent to which our conception of meals and food is being profoundly modified, echoing a way of living together that is fragmenting and recomposing itself into small real and virtual ideological circles.

The traditional meal that sealed the nuclear family and synchronized the home with neighboring households is gradually disintegrating as individualization of rhythms and digital nomadism become more powerful. We are seeing a similar pattern in many areas: the shift away from TV to targeted and personalized digital media, the erosion of the mainstream and the rise of more niche propositions in health, hygiene and beauty.

Evolution of Social Ties Around New Food Practices,
coordinated by Gilles SÉRÉ DE LANAUZE.

It seems that, in this period, what divides becomes more valued than what brings people together, even in the last bastion of living together: the meal.

Division, because today there is no more consensus about the direction of progress since this remains unresolved in the face of the ecological and economic crisis that had plagued our future even before Covid-19. From then on, several ideological currents have been fighting for the reins of progress to correct its trajectory. Degrowth and alternative thinking, affirmation and identity-based withdrawal, or belief in all-powerful technology, it is now necessary to choose our camp, our society, our future, our values; otherwise, a competing future might arise.

As this collective work demonstrates, such confrontation extends to our plates: eating is becoming an increasingly ideological and identity-based act, perhaps like all consumption. Qualified as vegetarian, vegan, religious, spiritual, hygienist, performist, meals and ways of eating increasingly divide the usual collectives and recompose networks that follow the divergent forces of today's ideologies.

This is an important statement with far-reaching consequences: first, because it announces the ambivalence of the role of the meal, which becomes as much an act of secession as of gathering. Tell me what you eat, I will tell you who you are not. Or tell me what you refuse to eat, and I will tell you who you are fighting. Then, of course, because it is changing the entire food universe, opening up new offers in new ecosystems for a total food experience: political, communal, but also esthetically pleasing, i.e. *Instagrammable*.

Finally, it makes you think beyond food, by reflecting one of the great paradoxes of our time in a stressful and isolating world: the contradictory need to join AND escape the collective.

In a recent survey for the World Economic Forum (September 2020), Ipsos notes that 72% of citizens in 15 countries clearly want the post-Covid-19 world to be decidedly different from the one before. But will we be able to all agree on this difference when our capacity to accept and manage otherness has probably never been so low? Or are we heading for competing microworlds?

Will the next generations be able to resolve this paradox, starting with the subject of food, which we know is a cultural pillar of Generation Y? How will the Covid-19 crisis influence these food mutations? Will it accelerate our food identity by reducing opportunities to eat together? Or will it, on the contrary, encourage a resurgence of the need to be together when the lockdowns are over?

So many questions that make this book essential for those who want to understand and anticipate the current evolution of social phenomena around food.

July 2021

Acknowledgments

This book was supported by the Montpellier Research in Management laboratory of the University of Montpellier (MRM-Axe Agro-Alimentaire) in partnership with the UMR MoISA of Montpellier.

All the co-authors would like to warmly thank Professor Jean-Marc Ferrandi for his precious advice, help and support in the editing process of this book.

Evolution of Social Ties Around New Food Practices,
coordinated by Gilles Séré de Lanauze.

Author Biographies

Laurie Balbo is a teacher-researcher at the Grenoble École de Management. Her work is mainly in the field of Social Marketing and, more precisely, on the study of factors that can affect consumer eating behavior.

René Pierre Beylier holds a doctorate in management sciences and teaches at the IPAG in Nice and at the universities of Montpellier and Avignon. A member of the MRM Laboratory (Montpellier), his work focuses on the territorial legitimacy of retailers and on consumer behavior, particularly with regard to local brands and products.

Foued Cheriet is a lecturer in international strategy and food marketing at the Institut Agro-Montpellier SupAgro and a member of UMR MoISA. He conducts research on the strategies of agribusinesses, agricultural sectors and entrepreneurship. Some of his work focuses on ethnic consumption (diaspora/dates, halal meat).

Margot Dyen is a lecturer at the University of Savoie Mont Blanc and a member of the Institut de Recherche en Gestion et en Économie (IREGE). Her work focuses on healthy and sustainable food practices and on well-being, in a social marketing perspective.

Fatiha Fort is a professor of marketing and consumer behavior at the Institut Agro-Montpellier SupAgro. She is a researcher at UMR MoISA where she conducts research on the marketing of local and regional products and on sustainable food consumption.

Steffie Gallin is a doctor of management sciences and a teacher-researcher at Montpellier Business School. Her research works focus on eating behaviors, social networks, online communities and social influence.

Evolution of Social Ties Around New Food Practices,
coordinated by Gilles SÉRÉ DE LANAUZE.

Andréa Gourmelen is a lecturer at the University of Montpellier (IUT Montpellier-Sète) and a member of the the MRM laboratory (Montpellier). Her work focuses on lifelong nutrition (students and seniors) and its evolution over time in relation to life events.

Félix Jourdan is a doctoral student in sociology at INRA in Montpellier. His thesis work focuses on the issues surrounding religious ritual slaughter in slaughterhouses. More broadly, Jourdan is conducting research on the evolution of norms, representations and practices of livestock killing.

Guillaume Le Borgne is a lecturer at the University of Savoie Mont Blanc and a member of the Institut de Recherche en Gestion et en Économie (IREGE). His research focuses on food waste and nutritional choices.

Marie-Christine Lichtlé is a university professor in management sciences (marketing specialization) at the University of Montpellier. Her research areas are communication, consumer behavior and social marketing. Her recent work focuses on food behavior.

M'hamed Merdji has been a full professor at Montpellier Business School since 1999. His expertise is mainly in surveys, data analysis, market research and strategic marketing at national and international levels.

Thibaut Nguyen is in charge of the Trends & Prospective Department within Ipsos, the French research institute. As a futurologist, Thibaut helps private and public players understand and anticipate the key dynamics that are shaping our society, impacting the way we will think, act and consume in the future.

Andry Ramaroson is a lecturer at the Centre Universitaire de Mayotte. His work focuses on the analysis of individual behaviors (household, business, etc.), entrepreneurship and innovation, and business performance. He is a member of the CHROME (Nîmes) and MRM (Montpellier) Laboratories.

Gilles Séré de Lanauze is a professor at the University of Montpellier. He is in charge of the agri-food arm of the MRM Laboratory (Montpellier). He conducts research in consumer behavior on attitude/behavior gaps and responsible consumption.

Lucie Sirieix is a member of UMR MoISA and professor at the Institut Agro Montpellier SupaAgro where she teaches marketing and food consumption. Her research focuses on consumer relations to sustainability, with a social marketing approach.

Erick Suarez-Dominguez is a doctoral student in management sciences, specializing in marketing, at the University of Montpellier. His thesis work focuses on the effects of social influences on vegetarian and vegan consumption behaviors.

Introduction

Eating Together, What Are We Talking About? Social Evolution of Today's Food...

Gilles SÉRÉ DE LANAUZE
MRM Laboratory, University of Montpellier, France

Can we still eat together? The question posed by this book may come as a surprise, especially in France, where the food model is still overwhelmingly based on a traditional organization of three daily meals, as well as on sharing meals with other diners, at 84% for dinner and 75% for lunch. However, while Rozin (1994) has shown that in the imagination of individuals, eating means sharing with others, around a table; this seems more difficult today, and we observe that for the last 20 years, the number of people eating alone is increasing, as well as the number of people eating outside the home[1]. Sharing the same meal around a table is becoming complex because of distance and the different agendas, concerns and paces of life of people who are close. Societal developments such as the rise of individualism, the break-up of families, or nomadism generated by the constraints of active life, weaken the traditional links that used to be formed around meals. Several ways of not "eating together" are thus revealed: eating alone of course, as well as not eating

1. See *Baromètre santé nutrition*, available at: http://www.ireps-picardie.fr/News/News_Cres_OR2S/Newsletters2010/17newsavril2010/Barometre_nutrition.pdf.

Evolution of Social Ties Around New Food Practices,
coordinated by Gilles SÉRÉ DE LANAUZE.

the same thing, not eating in the same place, eating within a different and chosen community. Thus, beyond the fact that we increasingly eat alone, many other fault lines seem to appear in the practices and social perimeters of food consumption. At the same time, other links are being re-formed within new communities of practice or belief, and the implementation of the shared meal depends on new forms of daily organization. And if we do not eat together as much as we used to, we may eat more connected to others, and with more conviction.

Sharing the same meal around a table also becomes complicated when food choices tend to be individualized. Exclusionary diets (gluten-free, lactose-free, meat-free, etc.) are multiplying. The development of intolerances and allergies to certain food products leads to a departure from the traditional patterns of gastronomy and conviviality. At the same time, growing concerns for health and the environment are generating new practices, new product offerings and new forms of market segmentation. Some practices are even more committed, such as veganism which, beyond the refusal of any consumption involving animal exploitation, often takes a stance of both societal disruption and proselytism. Thus, multiple factors are likely to generate differences, antagonisms or frictions in the relations and practices related to food. The development of particular diets, motivated by health, environmental or ethical concerns or imperatives, seems to generate new barriers to eating together.

Religious prohibitions are also at the origin of withdrawal into particular ethnic or community regimes. We also note that the religious issue has led to the appearance of a wide range of new products on the market (numerous halal and kosher product lines), which find their clientele in the corresponding community, as well as among other consumer segments in accordance with trends observed towards ethnic or cross-cultural products.

These changes in terms of relationships with others, conviviality and sharing go beyond the simple perimeter of the meals themselves and also characterize the practices associated both upstream and downstream from the meal. Upstream, we can mention shared supplies, either between consumers (grouped purchases) or between consumers and producers (AMAP[2], organic baskets), or the development of "cooking together" practices (courses, tastings, tutorials, reality shows, chefs' blogs). Downstream, we can observe the growing inclusion of gastronomy and commensality within other forms of sharing: meals are then offered not in classic

2. A French acronym designating the Association for the Preservation of Local Farming. It usually involves a partnership between a group of consumers and a local farmer.

and commercial places like restaurants, but in new alternative, tourist, campaigner or artistic contexts that rely on food to enrich other types of experiences. Commensality is then a pretext for sharing new experiences, emotions or convictions that go beyond their sole nutritional and hedonic objectives.

Sharing values rather than a meal: in an often anxiety-provoking context, consumers are looking for answers to the many economic, social, health and environmental issues linked to our consumption patterns; as a consequence, they are questioning the models of food consumption in order to take greater responsibility. Food takes on the dimension of supporting values. Societal or environmental convictions lead to the critique of certain categories of products deemed harmful because of their ecological footprint or their consequences on public health (meat, processed products, etc.). This favors particular modes of consumption, increased social pressure on what is good or bad to consume and retreating into communities of practice around new principles (organic consumption, local consumption, direct distribution, responsible consumption, etc.). Are these communities the new form of socialization around food that, in many cases, would replace the traditional family and social sharing?

While food is a means of supporting individual and social identities, it is at the heart of the relationships that we share with others. Food practices and convictions result in a chosen membership of groups, to which we identify or refer to, or on the contrary, from which we distinguish ourselves. These groups are an important element of the individual's social environment, whether physical or virtual, made up of relatives or strangers. They influence behavior and consumption practices and structure our identity. They play a role in the construction of the self, through adherence to and identification with models, or even communities of adoption, while at the same time, relationships outside of groups allow us to assert our own difference within society. Such communities are developing; they are sharing communities (Internet communities, religious or local communities) that are structured fully or incidentally around food. What are the mechanisms of adhesion to these communities, what returns do individuals expect from them and how does the community itself influence food behaviors? This question of social influences and intra- and extra-group relations on food is at the heart of this book.

Finally, these consumer trends are obviously reflected at the level of supply. Product innovation, distribution structures and the very organization of markets and chains of actors are accompanying these changes and new ways of eating. This is the case, for example, of the halal meat production chain, which has benefited from a complete reorganization of the production, control and distribution chains, with the emergence of new players and the development of an offer that takes into account both market expansion and identity support. Another illustration is the market for vegetarian products and services (catering, advice and support, literature), which is committed to the management of facilitated relations between followers and non-adherents. And we find new ways of consuming, as well as of producing and living together at the local level around food, with the re-enchantment of the terroir[3] as a place of sharing and conviction.

Between social influences, norms, communities and questions of identity, the objective of this book is to bring a *reflection on the current evolution of social phenomena around food and their influences on the evolution of these markets*. This reflection is articulated through six chapters briefly summarized as follows.

The recognition of the benefits of "eating together" is such that it has led to official recommendations from public authorities. This is the subject of **Chapter 1**, *"Eating Together", a PNNS Recommendation. How Can it be Put Into Practice?* In order to promote a balanced diet on a daily basis, the French National Nutrition and Health Plan (*Plan National Nutrition Santé*, PNNS) is based on the social aspects of eating: eating together on a daily basis would be a guarantee of a more nutritional diet, because we would take more time to eat, we would cook more recommended products such as fruits and vegetables, and this would limit snacking. However, as with other recommendations made by public authorities (five fruits and vegetables a day), if consumers seem to know the recommendation to "eat in company", its application is not satisfactory. It is therefore with the aim of overcoming this lack of application that this chapter focuses on the implementation of the "eat with company" recommendation. Data were collected using a qualitative study of food practices combining a projective collage method with individual interviews and observations. The results show that eating in company implies taking on a social role during meals and that culinary knowledge is used to satisfy the guests. The results also show that it is necessary to articulate our activities and those of others in time and space in order to meet around a meal.

3. "Terroir" is a French term that derives from the Latin *terra*, meaning earth, land or soil. In this book, we use this term as there is no direct English translation. A terroir refers to "a delimited geographical area defined by a human community that has built up, over the course of its history, a set of distinctive cultural traits, knowledge and practices, based on a system of interactions between the natural environment and human factors. The know-how brought into play reveals an originality, confers a typicality and allows recognition for the products or services originating from this space and therefore for the people who live there" (Casabianca *et al.* 2006)". See Chapter 6 for further details on this term and its usage.

Eating together thus implies grouping together around common practices, concerns and objectives, in particular by taking advantage of the opportunities for sharing and exchange provided by the Internet and social networks. The objective of **Chapter 2**, *"Eating Together" Through the Internet: The Case of Online Weight Loss Support Communities*" is to show how users of online communities "eat together" by sharing recipes, photos of the dishes they have made or even tips and advice on nutrition. This sharing contributes to the diversification of their diet. The study particularly explores online weight loss support communities such as Weight Watchers or forums on Doctissimo and Aufeminin in which diet is at the heart of discussions. Bagozzi and Dholakia (2002, p. 3) define online communities as "social spaces in the digital environment that allow groups to form and be maintained primarily through ongoing communication processes". The chapter focuses on the phenomena of social influences that take place in these communities (How do individuals see their food choices influenced by others?) based on rich and varied collected material. We analyze excerpts from discussions within communities, screen captures that highlight these exchanges, as well as individual interviews, partly conducted with nutrition experts (23) and partly with users of online weight loss support communities (25).

What are the psychological mechanisms that encourage sharing values rather than meals? Concern for the future consequences of our diet is one possible explanation, for example, in the case of individuals on diets. **Chapter 3**, *"Eating Together": With or Without the Dietary Constraints of Others?* shows how forced or chosen dietary restrictions can constrain "eating together". Firstly, the study highlights the distinction between the chronically ill and the healthy, and the perceived consequences of an unbalanced diet which are very different in these two cases. As a result, a diet constrained by a chronic illness may lead to difficulties in eating with people who do not have these constraints. This first section identifies the variables that explain these difficulties in "eating together" as well as the consequences on eating behaviors when meals are taken together. In a second step, the analysis is extended to "chosen" eating, through the notion of consideration of future consequences. Consideration of future consequences (CFC) is defined as an individual's consideration of the potential and distant consequences of his or her current behaviors and the fact that these behaviors are influenced by the very anticipation of these potential consequences (Strathman *et al.* 1994). In the case of eating, these are individuals who impose "dietary constraints" on themselves because they are aware of the consequences their eating may have on them or others.

This results in food restriction behaviors on their part that depend on the consequences they take into account (e.g. veganism, organic food, weight loss diet). Based on these observations, why and how does a person go against his or her dietary principles in a situation where he or she has to eat with others? Similarly, to what extent will a person who does not impose dietary restrictions on himself/herself bend to the dietary constraints of others when eating together? What about the perceived tolerance of the other towards ourselves? Through illustrations drawn from three focus groups, made up of about 30 people with diversified diets, this chapter provides some answers to these questions.

We thus witness a reconfiguration of the social environment of the food practice and adopting or retreating into the sharing community. This is particularly observable in vegetarianism and veganism, where social influences play an important role. **Chapter 4**, *Eating Together, Yes, But Without Meat! Social Influences Related to Vegetarianism and Veganism* analyzes this strong trend of vegetarianism today. This restrictive diet in a society that is still predominantly meat-based generally breaks with previous habits, with the probable consequence of generating numerous psychological and relational tensions for the people who adopt it. Beyond the practices and motivations that have already been widely documented, the purpose of this research is to identify the multiple influences leading to the adoption and maintenance of this diet and to better understand the relationships that surround them and the new adopted community. An exploratory study was conducted through individual interviews with 25 young adult vegetarians. The results show that relationships with the community may have more effect in maintaining behavior than in the original adoption process and only indirectly facilitate relationships with the environment. This could have implications in terms of support, supply and communication regarding these practices.

This can also be studied at the level of religious communities for which community, socialization and food have always gone hand in hand. **Chapter 5**, *Eating Together and Differently: Halal Between Standardization and Segmentation* paints a picture of the production and consumption of halal meat in France by examining the dynamics of differentiation and standardization. The significant development of this market over the last two decades has led to important transformations in terms of production systems, logistics, communication and distribution. The involvement of the main actors of the food system and the emergence of specialized companies account for the coexistence of these two strategic dynamics of standardization and differentiation of supply, and respond to an increasingly complex demand. Based on the exploitation of secondary data and those from recent surveys on this sector, the aim here is to present the specificities of the halal meat market based on an analysis of slaughtering and processing practices upstream and distribution and consumption downstream. Assuming a hyper-segmentation of supply, the issues related to the ritual slaughter of animals and the different systems guaranteeing the "halal" character of meat are discussed, as well as the way in which the production stage is structured and organized between the main actors. The study of standardization–segmentation phenomena then highlights the economic, regulatory and production issues underlying the "transformation" of the living animal into a halal meat product identifiable by consumers. The perceptions of the latter as to the credibility of the certifications or the "halal" character of the products make it possible to identify the link between the two levels of analysis (Upstream–Downstream). Finally, this question of standardization/differentiation of supply is studied from the point of view of demand, its characteristics, its evolution and the way it is articulated with supply.

Finally, another dimension of socialization around food, geography, through the valorization of proximity and terroir, reveals another social perimeter, that of communities at the local level. **Chapter 6**, *From "Eating Together" to "Living Together Better", the Case of Local Products* addresses these links between local products and the local area, close links that are claimed as such, within a perimeter of sharing, and around signs of quality and values linked to the terroir, but which beyond that also qualify the territory. If the links between local products and territory illustrate the evolution of the values and representations attributed to these spaces for more than half a century, we cannot help but notice today the prevalence of local consumption, more convivial, more ethical, more responsible, more committed to the territories. The multiplication of short channels, "local" brands, "terroir" brands and territorial brands attest to this search for territorial anchoring in terms of proximity, territorial identity, belonging and attachment to the territory, often with the argument of better eating and health. If food plays a major role as a support for the construction and expression of identity, eating local food symbolizes belonging to a common cultural space, marking territorial roots, and *in fine* a form of living together better. Is this not the objective of the glocalized eater who is trying to achieve two complementary movements that modernization had made contradictory: to be attached to terroir, on the one hand, and to become globalized, on the other hand? In other words, the networked local and the conviviality of the terroir that favors the links between the place and the body. "I am what I eat" (Fischler 1990), on the one hand, and, on the other hand, the commitment of the digital consumer using global trends, which is expressed in their desire to give meaning to their consumption and the importance of doing something for ourselves, "I eat what is good for the body and the mind", promoting the convergence of the values of the collaborative economy and those of sustainable and local development towards a desire to consume differently and better, in search of meaning and authenticity. Small daily gestures that give the feeling of taking part in the changes, of being a "hummingbird consumer", connected, demanding and more enlightened in search of transparency, who wishes to change things by infusing new behaviors. Isn't the modern local entrepreneur a digitalized actor, and the post-modern client an informed consumer, rooted in the territory and connected to the world? Isn't local a responsible commitment, a guarantee of quality and origin, of eating better and healthier for ourselves, others and the planet, of eating connected?

This local focus has suddenly been placed at the heart of current events, along with the strengthening of relational ties around meals and questions about food issues, with the Covid-19 health crisis and the experience of lockdown. **Chapter 7**, *By Way of an Epilogue: "Eating Together" in the Time of Covid-19*, sheds light on what this current period has changed and contributed to the book's reflections and research conducted, for the most part, before the crisis broke out. Four main points structure what the health crisis has generated and revolutionized in our eating habits and in the social influences at play. This chapter first describes the main changes in our eating habits and practices. It shows how digital tools have allowed us to maintain the link and overcome the difficulties generated by physical confinement, as well as their limits. It explores the changes or reinforcements of convictions and principles of life following the awareness of health, societal and environmental issues provoked or reinforced by the crisis. Finally, it returns to the question of locality, social proximity and reinforced ties within groups or communities in response to the challenges imposed by the experiences of lockdown.

The **Conclusion** summarizes the contributions of the book in a cross-functional manner and proposes new avenues. The academic reflections and research presented aim to better understand the current evolution of socialization and the influence of communities on food practices, and to draw enlightening perspectives for the different actors, food companies or distributors, consumer associations and groups, or public authorities.

I.1. References

Bagozzi, R.P. and Dholakia, U.M. (2002). Intentional social action in virtual communities. *Journal of Interactive Marketing*, 16(2), 2–21.

Casabianca, F., Sylvander, B., Noël, Y., Béranger, C., Coulon, J.B., Giraud, G., Vincent, E. (2006). Terroir et typicité : propositions de définitions pour deux notions essentielles à l'appréhension des Indications Géographiques et du développement durable. *Actes du VIème Congrès International des Terroirs Viticoles*, 3–7.

Fischler, C. (1990). *L'Homnivore*. Odile Jacob, Paris.

Rozin, P. (1994). La magie sympathique. In *Manger magique : aliments sorciers, croyances comestibles*, Fischler, C. (ed.). Editions Autrement, Paris.

Strathman, A., Gleicher, F., Boninger, D.S., Edwards, C.S. (1994). The consideration of future consequences: Weighing immediate and distant outcomes of behavior. *Journal of Personality and Social Psychology*, 66(4), 742–775.

1

Eating Together, a PNNS Recommendation. How Can it be Put Into Practice?

Margot DYEN[1] and Lucie SIRIEIX[2]
[1] Institut de Recherche en Gestion et en Économie (IREGE), University of Savoie Mont Blanc, Chambéry, France
[2] UMR MoISA, Institut Agro-Montpellier SupAgro, France

1.1. Introduction

What is eating well? There are many answers to this seemingly simple question. First of all, we need to define what we mean by "eating well". Historically, the issue of health through food was, quite obviously, mainly dealt with by nutritionists, who had an almost exclusively metabolic and physiological approach to nutrition. This led to campaigns promoting a balanced diet, making individuals aware of the nature of the foods to be consumed according to their category (carbohydrates, fibers, proteins, etc.). This functional vision of food prevailed in France until the 1960s and was perpetuated in English-speaking countries, but the book *Manger: Français, Européens et Américains face à l'alimentation* (Fischler and Masson 2008) shows how, contrary to this vision, the French preferred to conceive food another way. Today, nutritionists are increasingly interested in the issues of social relations around food and highlight that they are determining factors in a balanced diet. Therefore, eating well does not only depend on what we eat, it is also taking the time to sit down, in the company of people who allow us to enjoy a good time. It is also about sharing a dish, about a family meal established at well-defined times. This change can be seen in the

Evolution of Social Ties Around New Food Practices,
coordinated by Gilles SÉRÉ DE LANAUZE.

recommendations that are made and the French National Nutrition and Health Plan (*Plan National Nutrition Santé*, PNNS). "Eat and Move" campaign devotes an entire section to this dimension. Thus, *eating together* is now part of the promotion of eating well, which is no longer just a balanced diet, but an "art of living". Eating together helps in regulating food intake and invites people to cook more and to take more time to eat, which in turn helps in better perceiving the signals of satiation. As mentioned above, this "art of living" is particularly central to French culture, whereas English-speaking societies have retained a vision that is focused on the nutritional characteristics of food.

In the context of this book, which asks the question "Can we still eat together?", this chapter is therefore not about all the recommendations of the PNNS but about a specific recommendation: *eating together*.

Eating together underlines the PNNS's concerns, as will be detailed in section 1.2. However, this recommendation is not supported by the PNNS. In order to better support these practices of *eating together*, it is therefore necessary to better understand them and to question the conditions of their implementation. To do this, we adopt a practice-based approach in this chapter to understand how individuals eat together: a qualitative study conducted with 23 participants is presented in section 1.3. Section 1.4 more specifically addresses the questions "what are the materials, skills and meanings associated with these practices?" Section 1.5 focuses on the interactions between these three elements – materials, necessary skills and associated meanings – to show that together they can give rise to particular practices or help overcome obstacles to *eating together*. On this basis, in section 1.6, we examine the links between *eating together* and well-being, and in section 1.7, we open up perspectives for accompanying or facilitating *eating together* practices.

1.2. *Eating together*, a recommendation of the National Nutrition and Health Plan

Since 2001, France has had a real public policy targeting food, through the *Plan National Nutrition Santé* (PNNS, French National Nutrition and Health Plan). This initiative, launched by Prime Minister François Fillon in December 2000, had as its general objective "to improve the health status of the population by acting on the major determinant represented by nutrition" (Hercberg 2006). The PNNS materials include food guides, poster campaigns, press articles, support documents and television campaigns. Most of the communication tools are aimed at the general public, but some of the productions are intended for health professionals. The PNNS is intended to be applied in all areas of food consumption, with interventions in schools, health systems, the workplace, neighborhoods and cities, as well as in collective catering. The overall goal of the PNNS is to "contribute to the creation of an overall nutritional environment, facilitating positive choice for the health of consumers" (Chauliac 2011). As a public policy, the PNNS is equipped with evaluation systems to assess the effects of its actions. Thus, it "integrates the regular evaluation of its quantified objectives and, as much as possible, of the actions or measures it implements (effectiveness indicators or process evaluation)". Its objectives can be summarized in four main points:

– reduce obesity and excess weight among the population;

– increase physical activity and decrease sedentary behavior at all ages;

– improve dietary practices and nutritional intakes, particularly in at-risk populations;

– reduce the prevalence of nutritional diseases.

An assessment of the achievement of these objectives was carried out in 2006–2007 via the *Étude Nationale Nutrition Santé* (Castetbon *et al.* 2011), supplemented by the *Étude Individuelle Nationale des Consommations Alimentaires* (INCA2, National Individual Food Consumption Study (Afssa 2009)), the *Baromètre santé-nutrition* (Health and Nutrition Barometer) and regional studies (HCSP 2010). These assessments revealed that the objectives had only partially been achieved and that the program needed to be strengthened to continue to promote the desired behaviors (e.g. only a very slight increase in the purchase of fruits and vegetables had been observed) (Hercberg 2006). From this observation, the PNNS2 (2006–2010) emerged.

At the end of the PNNS2, the evaluation report noted that the expected results had not been obtained (particularly in terms of obesity screening (IGAS 2010)). This report recommends, among other things, revising the feasibility of certain objectives, correcting the formulation of certain principles that are sometimes the source of a binary vision of nutrition ("good" vs. "bad" products) and systematically citing the programs with which the PNNS is linked for better coherence between the public policy programs (Menninger *et al.* 2010). The PNNS is also criticized for not taking sufficient account of local specificities or social groups, which risks stigmatizing certain sectors of the population (Poulain 2006). For example, the PNNS3 includes a special section on overseas territories and emphasizes the importance of social, economic and cultural determinants of diet. This version explicitly emphasizes that it aims to banish "any stigmatization of people based on a particular dietary behavior or nutritional status" from the PNNS (Ministère du travail, de l'emploi et de la santé 2010). The PNNS3 also ensures overall coherence with other national food programs (such as the *Programme National pour l'Alimentation* (French National Food Program) resulting from the law on the modernization of agriculture and fishing, as well as the Obesity Plan, of which the PNNS is a component (Chauliac 2011)).

From its first version, the ambition of the PNNS was to "take into account the biological, symbolic and social dimensions of the food act" (Chauliac 2011). Thus, "eating well", a concept that is more encompassing than "nutritionally healthy eating", takes into account the social dimensions of eating and includes *eating together* in its program. However, *eating together* is not explicitly addressed in the evaluation. It is mentioned in the general parts of the reports, such as the preface, the foundations or the objectives of the program, but is not the subject of any explicit recommendations in the same way as there are consumption guidelines for fruit and vegetables, dairy products, etc. Despite this, on the *Manger Bouger* website, a section asks the question "Why is it important to get together for regular meals?"[1]. *Eating together* is therefore a more comprehensive practice of balanced nutrition, and is more in line with a balanced lifestyle. Indeed, *eating together* is beneficial for several reasons:

– promoting the structuring of meals (starter–main course–dessert), which itself promotes nutritional diversity;

– avoiding skipping meals;

– taking the time to eat (favorable to better digestion and a better perception of the signs of satiation);

– cooking fresh products (healthier than processed products, which are generally saltier, more fatty and less natural).

In France, despite the fact that the conviviality of meals remains central among the discourse of the French population (Mathe *et al.* 2014), *eating together* is less and less in line with the evolution of lifestyles, affected in particular by demographic changes (e.g. composition of households, activities of their members) or changes in food supply (e.g. prepared and individualized dishes). The INCA2 study published in 2009, based on data collected in 2006 and 2007, highlights a trend towards a destructuring of eating patterns among the youngest members of society (15–35 years), which has been increasing since 1999.

Similarly, it is clear that the time spent in front of screens and the solicitations of smartphones at all hours of the day and night can interfere with maintaining time to *eat together*, parasiting the interpersonal relationships essential to conviviality. The INCA3 study (Anses 2017) emphasizes that the time spent daily in front of screens (excluding work time) is constantly increasing, with an average increase over the last 7 years of 20 minutes in children and 1 hour and 20 minutes in adults. As a result, *eating together* is not self-evident, and more support is needed.

1. Available at: https://www.mangerbouger.fr/Manger-mieux/Que-veut-dire-bien-manger/Pourquoi-est-il-important-de-se-reunir-autour-de-repas-reguliers.

In conclusion, *eating together* is at the heart of the PNNS's concerns but is not the subject of accompanying measures, and current lifestyles make it difficult to maintain this practice which contributes to a healthy lifestyle. Faced with this observation, a study of the different practices of *eating together* and a reflection on the conditions of implementation of these practices are necessary.

1.3. Understanding the emergence and maintenance of eating together

This section adopts a practice-based approach to study *eating together* from a holistic perspective, that is, by taking into account all the elements related to the existence of practices. These elements are the material conditions, the necessary skills and the meanings associated with the emergence or maintenance of the practice.

1.3.1. *Benefits of practice theories to the study of eating together*

Practice theories are a set of theories and concepts whose principle is understanding everyday and routine activities by studying the emergence of practices as a result of the context rather than by analyzing the choices made by individuals to arrive at a behavior. Schatzki, a founding author of this theoretical trend, inherited from the work of Bourdieu and Giddens, defines practices as sets of activities that involve and are linked by knowledge, skills, meanings and material elements (Schatzki 1997). Reckwitz, another seminal author, positions himself in relation to Schatzki's ideas by adding the study of habits and routinized actions. Reckwitz (2002) defines practices as a type of routine behavior that consists of several interconnected elements: forms of bodily activities, mental activities, "things" and their use, accumulated knowledge in the form of understandings, skills, emotional states and motivational knowledge. Thus, social practices are taken as the unit of analysis; individuals are considered to be "carriers" of practices.

Shove *et al.* (2012), building on the seminal work of Reckwitz (2002), propose a simplified framework to empirically explain the emergence, maintenance and demise of practices. According to this framework, a practice emerges and is maintained as a result of a combination of three elements: skills, meanings and materials (Shove *et al.* 2012) (Figure 1.1).

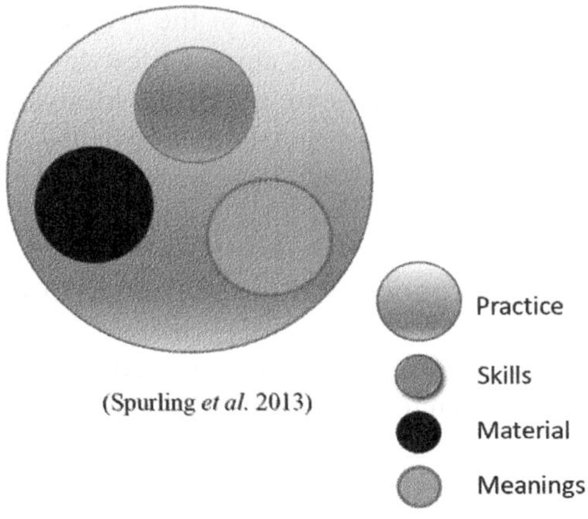

Figure 1.1. *Diagram of the conceptual framework from Shove* et al. *(2012), as presented in Spurling* et al. *(2013). For a color version of this figure, see www.iste.co.uk/seredelanauze/evolution.zip*

"Skills" refer to what an individual needs to know in order for a practice to emerge and be sustained. Skills are also found, in a more abstract way, in the knowledge related to a practice. Thus, under the term "skills", we can distinguish between physical abilities, know-how and knowledge.

The "material" element refers to what is tangible and mechanically or physically constrains or enables an action. This element groups objects and technologies (Reckwitz 2002; Shove and Pantzar 2005; Warde 2005) with infrastructure (Hargreaves 2011).

"Meanings" refers to "the social and symbolic significance of participating [in an action] at any given time" (Shove *et al.* 2012). This notion can be translated into the concept of the teleoaffective structure described by Schatzki (2005, pp. 42–55) or in terms of purpose and emotion, a form of goal-oriented engagement (Warde 2005).

Practice theories have been used to understand the emergence or to accompany the implementation of more sustainable practices (e.g. Hand *et al.* (2005), who study the evolution of hygiene practices and Gram-Hanssen (2010) who studies heating practices in relation to energy savings). Practice theories can furthermore be used to understand the emergence, maintenance or disappearance of practices at different scales: at the scale of a society (e.g. Shove and Pantzar 2005), or at a more individual scale, to understand how a particular practice is implemented by a particular individual or household (e.g. the practice of recycling in households

(Fournier-Schill 2014)). Our chapter opts for this second approach in order to understand, at a micro level, what explains the existence and maintenance of the practice of *eating together* at the household level. The following section presents a qualitative study conducted for this purpose.

1.3.2. *A two-stage qualitative study to understand how consumers "eat together"*

The methodological challenge of our work is not so much in observation, or in revealing the existence and arrangement of practices, but rather in understanding their origin or their construction. Moreover, applying practice theories does represent a methodological challenge, as it can be tricky to accurately describe how an action is socially constructed (Hargreaves 2011). Our work is therefore exploratory and comprehensive in nature, which justifies the choice of a qualitative approach. The research methodology is composed of two stages, which serve to describe and understand the implementation of the "eating together" recommendation by consumers, in order to identify the skills, material and meanings on which it is based.

1.3.2.1. *Step 1: Semi-structured interviews based on projective collages*

The qualitative approach involves looking for diversity of practices, with the aim of understanding the mechanisms of adoption of these varied practices. Turrentine and Kurani (2007) propose a sampling method known as "illustrative sampling" which aims to ensure the diversity of the object studied. As our qualitative approach involves looking for diversity of practices, the first step of this sampling method consists of studying the determinants of the phenomenon (socio-demographic, motivations, etc.) in order to know on "which diversity" it is most relevant to base the selection of the individuals in the sample. For example, it has been shown that which socio-professional category an individual belongs to conditions their involvement in food (Plessz and Gojard 2014), which can impact the importance given to meals. Similarly, it is relevant to choose households of different "types" (e.g. a couple, friends sharing a flat, family with children, etc.), which seem to generate different practices (CREDOC 2017), just like the fact that the nature of the relationship with people at the table conditions food intake (De Castro 1994). Thus, we selected households of varying socio-professional categorues and size and where relationships vary: family relationships, roommates, etc. It should be noted that the approach does not aim to compare households with each other, but to determine which characteristics should be used to generate the most variability. The sample thus guarantees a variety of practices with respect to *eating together* by varying the age, the number of people in the home and whether or not there are children.

Twenty-three participants were recruited via an Internet mailing list (Appendix, section 1.8).

The projective collage method consists of asking individuals to paste together images to represent what they think or imagine about a situation or concept. Projective methods have been developed primarily as a dialogue facilitator. They are generally used to reduce bias, to study individual perceptions and to access unconscious processes or mental images in consumers. Collage allows consumers to express themselves without introducing bias.

In this study, the collage was used to provide respondents with a self-constructed discussion aid in a time prior to the interview, to address a topic (the organization of daily *eating together*) which can be tricky to transcribe in an impromptu discussion. The poster was not analyzed as such.

To make the poster, participants were given a magazine. The participants were given the following instructions:

> Make a poster that represents your food practices: from shopping to consumption, and represent what conditions your food organization. You have at your disposal magazines, cutting and pasting materials, and drawing materials. You have as much time as you want for this workshop. Once completed, we will discuss this poster, it will not be interpreted without your help.

The collages were made individually by each of the 23 study participants. Participants were only informed that the study was about eating practices, so as not to focus their attention on the issue of *eating together*.

Participants took an average of 40 minutes to complete the exercise (e.g. Figure 1.2).

The collage method allowed us to recall and develop detailed descriptions of the individuals' daily activities, and to know the links that connect them.

Following this creative phase, an interview was conducted with the support of the poster. The interviews were recorded and transcribed. Non-targeted thematic coding was conducted in an exploratory approach. The coding was carried out by continuous thematization, which amounts to constituting the themes as the interviews were analyzed, which were taken in the alphabetical order of the participants' names. The analysis was based on this coding in order to identify the different practices of *eating together* and to characterize the elements necessary for their implementation.

In order to understand the context in which the described practices are embedded, the projective/discursive collection is combined with an observed/discursive data collection, as presented below.

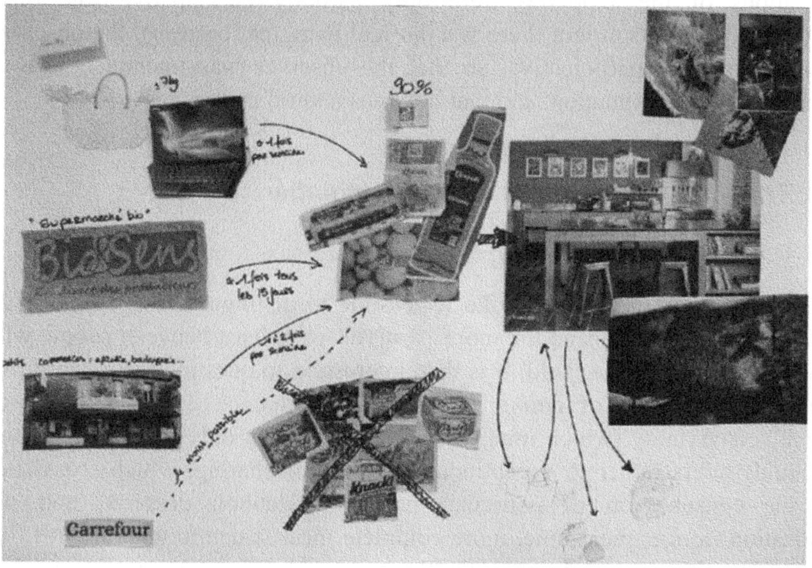

Figure 1.2. *Example of a projective collage (Paola). For a color version of this figure, see www.iste.co.uk/seredelanauze/evolution.zip*

1.3.2.2. *Step 2: In-store and at-home observations*

Approximately one year after the collages/interviews (the length of time that allowed for analysis of the collages/interviews), observations/interviews were conducted during 1) grocery shopping and 2) a meal in the home and its preparation and storage. 11 of the 23 participants from Phase 1 took part in this phase. These observations, although one-offs, made it possible to objectify the facts related in the first interviews (e.g. to see what a participant meant by: "my kitchen is very small") and provided a contextualized overview of what was described in the first interviews. Ideally, the interviewer visited the participants' homes before they left to go shopping, so that they could travel from home to the grocery store with them. In cases where it was not possible to meet at home, the meeting took place directly at the store.

The second phase of observations took place in the home, to collect data on the preparation of a meal, its progress and the tidying up that follows. During this phase, the objective was, in addition to obtaining additional information on the sequence of actions and the way they were carried out, to observe the interactions between the members of the household. All the members of the household had to be present during these observations. The observations lasted from two to five hours. If the observations of the shopping were only partially participative (the observer accompanied the person, but there was no real participation in it), the observations of the meals were participative, so that the observer was integrated as well as possible in the daily situation, and that it was as natural as possible.

1.3.3. *The different practices of eating together*

1.3.3.1. *A wide variety of eating together practices*

To begin with, the results make it possible to distinguish two types of *eating together*: the "everyday" and the more "festive", or at least between people who are not part of the same household. It is then interesting to question the extent to which these different forms of *eating together* can influence eating well. So-called "festive" conviviality (which implies that we do not eat only with members of our household) refers rather to exceptional moments of sharing, which are associated with the consumption of particular products (alcohol, desserts) and special organization (longer meal time, more complete meal structure with aperitif, starter, main course, cheese, dessert, etc.). On the host's side, there are unspoken rules: the products to be favored are rather rich products and they must be provided in ample quantities. For the guests, one rule is that they must conform to the meal that is offered to them and that hunger must take a back seat in order to honor what has been prepared for them. Camille explains:

> *– So you can estimate your hunger level, but during meals you sometimes force yourself to sit at the table and say "I'd better not leave this".*
>
> – No, we never get hungry, we put it in the fridge and we know we'll eat it at any time, no, I never force myself, at home anyway [laughs].
>
> *– Do you force yourself when sitting at other people's tables?*
>
> – Well, yes, if you have to.
>
> *– Why if you have to?*

– Well, I don't know the kitchen, there's… I know that grandma serves you and if you don't finish the plate she gets the impression that it's not good, it's very stupid, but you know that you have to finish; but it doesn't happen to me often.

– *You don't want to offend?*

– Yes [laughs], yes, I don't want to offend grandma!

These results refer to the fact that, under such conditions, individuals are more likely to eat beyond their hunger and, what is more, eat products with high energy density.

In the context of daily *eating together*, the practices of the individuals present at the table are modified. Indeed, the rules integrated by the person in charge of the food, in the role of a *good provider* (Evans 2011), instead assume the provision of varied and balanced food, while taking care to ensure the pleasure of the people eating. Roxane explains:

– We rarely eat the same thing twice in a row, it happens but hey…

– *Why?*

– Well, because it's true that at lunchtime… it's true that we like eating in the evening.

– *OK, you both feel like you eat fast at lunch so in the evening you don't want to do that?*

– Yes, that's it: it's a more quiet time.

Thus, the feeling of shared pleasure around a meal implies eating certain foods and avoiding others (e.g. leftovers). In particular, the results confirm that starting a household or having a child encourages the preparation of balanced meals.

In addition to these results, between an *eating together* practice that can be characterized as "festive" and another that is more "daily", the results make it possible to distinguish households in which *eating together* is established/systematized/codified, whereas in others this practice is more flexible/random. This is notably the case of Amélie, who tries to maintain the family meal with her daughter, despite the fact that she has difficulty extracting her from her video games:

> That's the problem, she spends a lot of time playing video games, so it has a lot of influence on our eating habits because it's often "come to the table, come to the table, come help me make food, save, come and cook dinner…!" So, in fact, I waste time asking her for help, so

> after a while I get tired of it, I leave what I'm doing, I give up, I make fries and ham, "come on, you're eating, you have to go back to school", so it has an influence.

For Noémie and her mother, on the other hand, the rules of the shared meal are established, knowing that "*every evening [they] eat together*", and that they have even gradually established weekly rituals such as "*leftover meals*" during which they finish all the leftovers of the week, as well as "*junk food meals*", during which they allow themselves to deviate from their respective diets.

In conclusion, the results allow us to account for the diversity of practices by distinguishing between *eating together* practices that can be described as "festive" or as "everyday", and by distinguishing between systematized/anchored *eating together* practices and others that are more flexible/random.

1.3.3.2. *Eating together: the influence of constraints and the centrality of food in the meal*

The practices of *eating together* can be distinguished according to the degree of constraint they represent for individuals, and according to the centrality of food during the shared meal. We thus identify *eating together* on a day-to-day basis, with the members of the household, but which can be experienced as a constraint or as a source of frustration.

> – She is absorbed by her video games, so it is very difficult to have habits, to stick to my eating habits and to impose them on my child because she is the child of divorce, there are habits that she has at her father's house and those she doesn't… that I find difficult afterwards when she comes back here to make her break the habit.
>
> *– But when she's here, do you eat at the table?*
>
> – Well, I eat by myself, so that's during the week when she goes to school, and on weekends we eat at the table (Amélie).

In this type of practice, consumers try to maintain *eating together*, either to follow nutritional recommendations or because (like Amélie) they see it as an opportunity to share a moment with their loved ones. For others, *eating together* on a day-to-day basis with household members can be synonymous with shared pleasure on a daily basis. In this case, it is generally a spontaneous and pleasant practice:

> It is also when we eat almost directly after we've arrived home from work, we cook and sit at the table, so it is still the time when we tell

> each other about our day, we are both a little quiet and we do not want to be disturbed by our phones (Marie).

In a more festive setting, *eating together* can be experienced as a way to get together with friends/family for special events (birthdays, reunions), but without this moment being centered on the food or on the cooking and esthetics of the meal:

> – *If someone comes to eat at your house, what do you do?*
>
> – Well it's… I have to force myself to anticipate I know that so-and-so is eating here so what am I going to prepare for them… and often I'll make them the simplest things possible because I hate elaborate dishes, so often it's going to be risotto with a salad as a starter.
>
> – *OK, so in fact when people come here it's just to have a good time, but eating together is incidental because you have to eat?*
>
> – Exactly (Sabrina).

On the other hand, in other cases, *eating together* allows friends and family to get together, but the food is central. Thus, the experience of *being together* sometimes begins even before the meal (guests come to cook):

> The last time we had a Mexican night, so we made fajitas and stuff, the other night we spent three hours making sushi, everyone with their own thing [laughs], so yeah, it's not bad, it's true that we cook a lot with our friends too, it's great (Camille).

In these types of practices, the esthetics and content of the meal are central.

Based on these findings, the remainder of this chapter offers an analysis, in the form of a discussion, of the conditions for the emergence of *eating together* practices. This analysis is conducted in light of Shove *et al.*'s (2012) framework of practice theories, described above.

1.4. Eating together: materials, meanings and skills

Eating together may seem like a harmless and simple practice for many people. For others, however, there are many reasons why this practice does not emerge or continue. The purpose of this section is to shed light from practice theories (and in particular from the theoretical framework of Shove *et al.* (2012)) on the question of the feasibility of *eating together*, with a view, in particular, to opening up avenues for facilitating *eating together* in everyday life.

Meeting around a meal, whether in a daily family setting or for more exceptional and/or convivial meals with people from outside the household, depends on the "material" feasibility of this meeting: having sufficient and suitable space, and being able to organize the physical meeting. Thus, there are material elements on several scales: on a more microscale, tangible elements are used to implement the practice: for example, having tables and chairs, plates and cutlery for the guests:

> Personally, I eat standing up at home in the corner of the kitchen, which is very bad, but as soon as I am in Montpellier [at my girlfriend's house], I absolutely have to take my tray, my plate and sit down at the table, it is a *sine qua non* condition, but otherwise she does not accept that we eat standing up near the corner of the fridge (Maurice).

Beyond the question, on a microscale, of the material needed to *eat together*, there is the question of the logistical feasibility of the meeting. This is a broader scale of the material environment, corresponding rather to the infrastructure. Everyday life is arranged in a given time and space, in which consumers carry out their practices. In this respect, *eating together* depends on the possibilities offered by the location of the place where people live, by the time it takes to travel from home to work, by the location of friends' homes, by the time it takes to travel between the place where they do their evening activities (gym, music lessons, etc.) and their homes, etc. These arrangements offer more or less the same opportunities for the consumer to do their daily activities, but varying degrees of opportunity for individuals to share meals with members of their own household or with friends/family from outside the household. It is interesting to note that this logistical feasibility of *eating together* is tending to evolve, particularly with the emergence of practices such as *remote commensality*: the fact of sharing a meal through a video call, made possible by new technologies and mobile applications: "I often call a friend, I know that he is eating alone so we call each other and eat together [laughs]" (Camille). Note that Camille's comments predate the Covid-19 pandemic. These practices, which do not limit *eating together* to physical encounters, probably flourished during (and perhaps since) the lockdowns that took place in France from March 2020 due to the Covid-19 pandemic. In any case, it is essential to note that *eating together* depends largely on the material conditions that allow individuals to meet and organize the meal.

In addition to the tangible material elements that underlie *eating together*, skills are also involved in implementing practices of *eating together*. Having the skills to choose, buy, store and prepare products is a first set of skills, which are broader than just cooking a meal. More broadly, cooking products that are adapted to the people with whom we are eating to make sharing possible between the commensals can be an issue. This sharing supposes that individuals can be gathered around a dish that will be accepted and appreciated by all. This essentially refers to the knowledge and skills of the person in charge of the organization and preparation of the meal, which consists of the management of the event itself, as well as of the food preferences and particularities of the guests, as well as the management of the products:

> From Monday to Saturday we eat together every evening, knowing that on Saturday there are two meals: at noon and in the evening, on Sunday at noon we eat with my grandmother, so there will be three of us at the table, and I will also adapt all the shopping etc. to Sunday, because my grandmother no longer has the same taste as we do, so she can't stand certain foods, whereas we love them, well, I adapt to what she likes because she has to eat, and she is the one who needs to eat the most (Noémie).

Similarly, beyond a question of taste, skills are needed to take into account considerations related to the household economy, so that the products consumed when *eating together* are in line with the household budget, which must be able to "afford it".

A second set of skills refers more to the skills and behaviors involved in eating. Respecting codes of good behavior during meals is culture-specific (e.g. eating with cutlery, waiting for commensals to finish the dish before moving on to dessert, etc.), and this set of skills is closely related to meanings, which is the third element that helps explain the emergence and maintenance of a practice according to Shove *et al.*'s (2012) theoretical framework.

Meanings can be thought of as the values necessary for a practice to emerge or be sustained. Thus, regarding *eating together*, we can assume that meanings are found in two aspects: 1) it is necessary for *eating together* to be part of the values of the individuals involved and 2) it is necessary for the values of the commensals to be reconcilable during a meal.

The first aspect assumes that mealtimes are maintained by individuals if they attribute a value to shared meals (or even a symbolism that can lead to ritualization). Without being exceptional or festive, daily family meals can have a central value in the life of a household, contributing to the maintenance of the practice of *eating together*:

> I noticed that when [my boyfriend] wasn't there, we weren't together for a week or two, I ate less, so I think that there is a social side to it, we like to have a meal together in the evening, you know, so I think that this brings a bit of a rhythm in the sense that we will try to agree

> to eat together, if we can, we don't do… it's not independently each of us on their own and we eat together, we organize ourselves a bit to have a meal together… (Paola).

The second aspect related to meanings more specifically concerns the time of the meal, its feasibility and its good progress. Indeed, the good progress of a meal with several people is not defined by all individuals in the same way, and the feasibility of *eating together* depends on this agreement: for some, *eating together* begins in the kitchen:

> – *Do you cook for the others every day?*
>
> – Yes, every day and we cook a little together with my roommate because she loves to cook, so we usually start with a base of foods that we want to eat and then we assemble a lot of things, one idea and then the other, and it creates something each time! (Camille).

For others, on the other hand, it is really the time spent together that matters, and the preparation/consumption of food is far from central:

> You see, for example, I have a friend who comes to eat and she knows how I work and when she comes to my house she is always the one who brings food because she knows that I get bored, so she says to me, "Here, you take care of this, this, this, and I'll take care of the food", so that's fun and it's true that my friends are starting to get to know me (Pauline).

Thus, the place given to food when *eating together* is one of the meanings that conditions a harmonious sharing of meals.

Similarly, another meaning that can condition the feasibility of a shared consumption time is the relationship that the commensals have with the fact of eating or not eating the same thing:

> Each person helps themselves, except if it is something that is a bit boring, so we serve that, but otherwise each person does it in the order they want because there is one person who wants salad like that and the other doesn't want it… everything is on the table and we help ourselves […]. So it's true that he eats green salad, he can't finish without green salad for example, and he'll eat a dessert, either yoghurt or something… a dessert (Roxane).

In this case, we can see that, for Roxane and her partner, *eating together* does not necessarily mean eating exactly the same thing. Conversely, Myriam feels "frustrated" that her partner does not have the same tastes as she does, and this taints their family practice of *eating together*: "– Cooking vegetables is easy. – *If you don't do it, it's because of other people's tastes, right?* – Yes, that's it, in fact I'm frustrated at home".

These questions are particularly topical nowadays, because of the individualization of food, with more and more special diets (gluten-free, vegetarian, lactose-free, etc.). For example, in order to respect the meanings associated with the family meal, some vegetarians make exceptions to their diet and eat meat. Paola bends her vegetarian diet during family meals to avoid being the target of questioning and to avoid being forced to justify herself. Thus, for her, the meeting with her companions takes precedence over her personal values: rather than not eating meat for the sake of animal welfare, she prefers to get along with her family during meals. Similarly, Camille explains that she never forces herself to eat, except when she is at other people's houses: "I know that grandma serves you and if you don't finish the plate she gets the impression that it's not good, it's very stupid, but you know that you have to finish, but it doesn't happen to me often" (Camille). In this case, sharing a meal means honoring the dish that has been prepared and served by the host.

1.5. Interactions between materials, meanings and skills: particular practices or means of overcoming constraints

The previous section presented the elements (material, meanings, skills) one after the other, in order to describe their nuances and specificities. However, according to the framework of practice theories, it is the combination of these three elements that allows a practice to exist. This section therefore illustrates cases of interaction between the elements: when the interaction between the elements allows for the emergence of particular practices (such as convivial festive consumption), and when the interaction between the elements allows for the removal of constraints to *eating together*.

Skills and meanings can be closely linked in a festive *eating together* practice, in which the guests (and usually the host) demonstrate specific skills, which are at the heart of the *eating together* practice: this is, for example, the case for meals in which the objective is to discover or showcase a recipe and the dish is a major topic of discussion during the meal. In the same way, the esthetics of the table stands at the border between the material (e.g. beautiful dishes) and the meanings: welcoming well implies "pulling out all the stops":

> – I'm a fan of plates, glasses, cups and all that, it's part of it, I love it, even for Elliott, I love children's things and all that, so it's part of the pleasure of the table.
>
> – *It goes to the point of making little presentations?* Yes, when we do... well, not every day, let's relax (laughs), but when we have friends over,
>
> – yes, I do it. Before I knew Eliott I used to do it a lot, but now [...] I do it a little less, I do it mostly when we have friends coming over, that's it, to make a nice table I like everything that is, contrary to this drawing, everything that is mismatched, mismatched plates, I love everything that is a little vintage, 60s, a lot of stuff in that style, Scandinavian stuff, that's it, I like everything related to crockery... (Roxane).

Thus, we see that the material can be seen as a basic constitutive element, but that there is a very wide range of material around *eating together*, in response to particular meanings and skills.

From the perspective of supporting the emergence or maintenance of *eating together*, it is also interesting to understand how the interactions between the elements that make up the practices make it possible to overcome certain constraints *eating together encounters*. Indeed, it may be relevant to target the elements according to this hierarchy. Material constraints, which can be considered central to the deployment of a practice, can, for example, be circumvented when the practice can call upon meanings and skills that are already in favor of *eating together*:

> – But you see, when we were in the castle, there must have been 15 of us, now there are a few less.
>
> – *And where do you do it?*
>
> – Well, in the living room.
>
> – *Do you have a table?*
>
> – Yes, but it's a low table, so we eat sitting on the floor with plates, but there you go (Marion).

On the other hand, material conditions that are in favor of *eating together* (e.g. sufficient dedicated space) can be ignored when meanings and skills do not support *eating together*. The driving forces or the obstacles to *eating together* seem to depend more on the meanings, in the face of which the skills are called upon to get around any material obstacles. This leads to the next section, which discusses the constraints that can work against *eating together*.

Organizational constraints can indeed complicate the fact of *eating together*. However, in some cases, organizational skills, carried by the meanings in favor of *eating together*, can counteract these constraints:

> Cecile has to get up at 6:00, I have to get up at 6:30… so she has time to do a bike session or a little bit of gym activity, something like that. Then she goes for a shower and I take over there on the bike, and when I go to wash up she takes over in the kitchen, usually to prepare something [for lunch and breakfast], and usually when I get out of the shower I give her the final bit of help, then we eat (Thibaut).

We can thus say that *eating together* is a complex practice that solicits the constituent elements of practices at various levels (Figure 1.3).

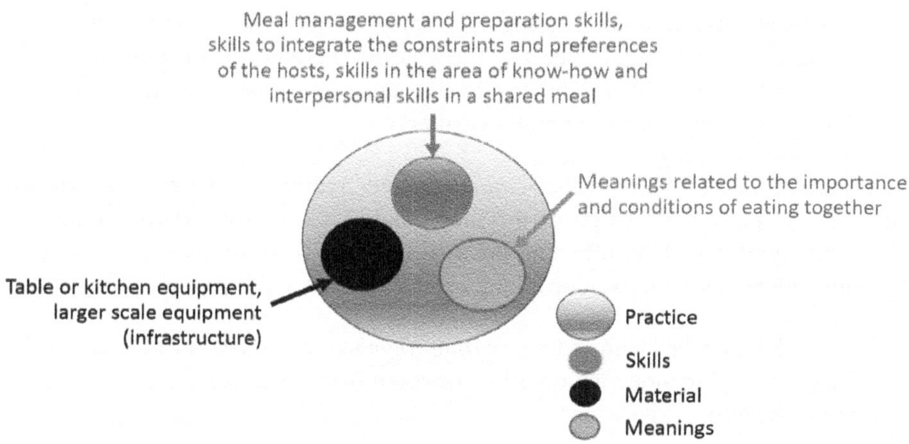

Figure 1.3. *Summary of the components involved in the implementation of eating together. For a color version of this figure, see www.iste.co.uk/seredelanauze/evolution.zip*

The interactions between these different elements sometimes make it possible to maintain the practice of *eating together* even when conditions are not optimal.

1.6. Does eating together always promote well-being?

As the previous sections show, the meanings associated with *eating together* are essential in maintaining this practice, especially when it is associated with a fulfilling and healthy family/social life:

> "It's still nice to share a meal" (Camille), "Eating alone quickly becomes a mess" (Marion).

These quotes summarize how conviviality is, quite intuitively, a positive aspect of eating. However, in other cases, *eating together* can be seen as a constraint:

> – *The constraint would be to do a lot, in fact?*
>
> – Yes, to do a lot and to do it in a very repetitive way every day, every day, every day [...]. I knew that in my family – because we were a large family – so I always remember my mother saying to herself, "What am I going to do tonight?" I know that I have things and so it's the desire to eat something at a given moment, well, I say to myself, "I want this, how am I going to prepare it" [...], that's why I put pleasure in the middle because there is the pleasure of touching the products, mixing them and the smells, the colors, the heat, etc., and then preparing something pretty (Esther).

Thus, this quote confirms that *eating together* requires logistics and upstream work for the person in charge of the meal, and that eating alone can sometimes relieve this burden without affecting the quality of the food. Moreover, it illustrates that eating alone does not exclude taking care of our diet and enjoying eating well.

Earlier, the chapter identified that *eating together* refers to a set of more or less tacit rules, and eating alone would allow freedom from them, which may, for some, be seen as a benefit:

> My diet has become more chilled, in the sense that the fact that living with my parents has changed the way I eat, because in a way we are infantilized when we live with our parents as adults and we are subjected to "don't do this, don't do that, we eat at such and such a time, we do this, we do that" (Bérengère).

This may allow the pursuit of certain objectives, which are more difficult to achieve during shared meals:

> I realized that I lose weight more easily when I am alone because I manage my rhythm, I let go, I am not managed anymore because of the look of others, not that of my daughter, nor that of my boyfriend, nor that of my family, I manage myself calmly, I am at ease and I have the impression that the weight goes away by itself too (Amélie).

These findings invite the question of well-being related to *eating together*, echoing the work of Mugel *et al.* (2019). This work employs the concept of eudemonic food well-being, according to which well-being is not only about pleasure (hedonic well-being) but also about the role of experience and the meaning given to actions. In light of this concept, we understand that *eating together* is not an absolute guarantee of well-being, and that it is necessary for the experience to be in line with the meanings of individuals for it to be a source of well-being. More broadly, particular circumstances, such as the lockdowns experienced by the French from March 2020, make it possible to qualify the fact that *eating together* is always a vector of well-being: in circumstances as constrained as the lockdowns were, preparing a meal every day for a household in which relationships are not always easy can be a major challenge, and the shared meal does not always have the power to generate conviviality and shared well-being. This invites consideration of a complex relationship between *eating together* and well-being that takes into account the constraints: *eating together* can increase well-being, but it is also necessary that some form of well-being (e.g. good understanding between guests) precedes – or is associated with – *eating together*.

1.7. What are the perspectives for promoting eating together?

The objective of this chapter was to better understand the conditions that influence the implementation of *eating together*, in order to reinforce the recommendations made in the PNNS. In particular, it allows us to distinguish the different practices of *eating together* (festive vs. daily, systematic vs. flexible). Moreover, this work shows that meanings play a major role in the emergence of *eating together* practices. In order to contribute to the maintenance or acquisition of a positive meaning for consumers, encouraging them to maintain this practice even when the material conditions are not necessarily present, it is essential to address all consumers, taking into account the various practices that exist: *eating together* to meet, *eating together* to exchange ideas about cooking and *eating together* in the face of different constraints or in different contexts. The practice-based approach allows for a good understanding of *eating together* according to the context, making it possible for several forms of *eating together* to be adopted by a given individual, as shown in the study presented.

In order to foster *eating together*, the public authorities could broaden the scope of the PNNS recommendations by not limiting themselves to the discourse on the website. A first step could be a real communication campaign, like the one undertaken for fruit and vegetable consumption. In connection with the results presented in this chapter concerning the importance of the meanings associated with *eating together*, such a communication would promote maintaining the institutionalized shared meal time in society. But communication alone cannot guarantee the emergence or maintenance of a practice (Keller *et al.* 2015); it requires concrete relays, particularly to counter changes in society that erode the practice of *eating together*. Thus, maintaining times and spaces dedicated to *eating together* would materially contribute to the implementation of this practice. We could thus ensure that meal breaks are maintained during the work day, and that workplaces are equipped with rooms adapted for shared meals (tables and chairs, microwave oven to bring our own food, etc.). Outside of the workplace, initiatives such as "*Voisins solidaires*" could provide an opportunity for people who are unable to eat with family and friends even if they want to. The association Neighborhood Solidarity (*Voisins solidaires*[2]), launched by the founders of the *Fête des Voisins* (part of www.world-neighbours-day.org), is developing a national mobilization scheme to strengthen local solidarity throughout the year. The actions are carried out through seasonal meetings (back to school, Christmas, etc.) and thematic meetings (extreme cold, waste sorting, etc.). We could imagine making these meetings even more regular and developing a theme targeting meal sharing in order to offer regular opportunities to *eat together*. Such a system could be the relay of the recommendations made by the public authorities within the framework of the PNNS, from a real perspective of accompanying consumers.

Other initiatives could also address the issue of screens, which are currently one of the major competitors of shared consumption times. In order to *eat together*, we must not be parasitized by screens, which capture attention to the detriment of the attention paid to the guests. The PNNS website recommends: "no screens at the table". To accompany this, phones could now be equipped with a "meal" mode like the "airplane" mode, which would allow for selecting the notifications to be filtered (e.g. we can keep the possibility of being called in case of emergency and receiving notifications on the news, but cut the notifications of social networks). This would also have a strong incentive dimension, legitimizing the fact that meal times should not be interfered with.

To conclude the reflections, this chapter also invites us to explore the new forms of conviviality and *eating together* that are emerging. This is especially true since

2. Available at: www.voisinssolidaires.fr.

the recent lockdowns due to the Covid-19 pandemic have probably redefined some of the basics of household food consumption. Moreover, this unprecedented situation has prompted many consumers to practice *cyber-eating together*: having "Skype happy hours", making "Zoom meals", etc. The materiality, skills and meanings of *eating together* are probably redefined in these situations. Beyond the new conditions for implementing the practice, these developments may raise questions in terms of promoting *eating together* for nutritional reasons: do we retain the nutritional benefits of commensality when screens are invited to the table?

1.8. Appendix: Sample summary

Respondent's code name	Gender	Structure of the home	Age
Marion	F	Lives with a partner	33
Esther	F	Lives alone	67
Paola	F	Lives with a partner	29
Valérie	F	Lives with husband and two children	48
Noémie	F	Mother	26
Bérengère	F	Lives alone	29
Arnold	M	Lives with wife and daughter	46
Maurice	M	Lives alone	69
Roxane	F	Lives with husband and son	33
Camille	F	Shares with a friend	21
Amélie	F	Shared custody of daughter	42
Coline	F	Mother	48
Myriam	F	Lives with husband and son	42
Sylvie	F	Lives alone	30
Justin	M	Shares with two friends	24
Sabrina	F	Lives alone	23
Vincent	M	Lives alone, although sometimes with his son	52
Léa	F	Lives with husband and two daughters	30
Thibaut	M	Lives with a partner	32
Marie	F	Lives with husband and newborn	32
Edouard	M	Lives with a partner	29
Cédric	M	Lives with wife and two children	37
Max	M	Lives with wife and two children	40

1.9. References

Afssa (2009). Étude individuelle nationale des consommations alimentaires 2 (INCA 2) 2006–2007. Report, Afssa.

Anses (2017). Étude individuelle nationale des consommations alimentaires 3 (INCA 3). Report, Anses.

Castetbon, K., Lafay, L., Volatier, J-L., Escalon, H., Delamaire, C., Chauliac, M., Ledésart, B., Hercberg, S. (2011). Le Programme National Nutrition Santé (PNNS) : bilan des études et résultats observés. *Cahiers de Nutrition et de Diététique, Programme National Nutrition Santé*, 46(2, Supplement 1), S11–S25.

Chauliac, M. (2011). PNNS : 2001-2006-2010-2015 Stratégies et dynamiques [Online]. Available at: http://inpes.santepubliquefrance.fr/jp/cr/pdf/2011/session7/Presentation-Chauliac.pdf [Accessed 20 October 2017].

CREDOC (2017). Comportements et attitudes des consommateurs à l'égard du commerce alimentaire en 2017. *Cahier de recherche*, 336.

De Castro, J.M. (1994). Family and friends produce greater social facilitation of food intake than other companions. *Physiology & Behavior*, 56(3), 445–455.

Evans, D. (2011). Blaming the consumer – once again: The social and material contexts of everyday food waste practices in some English households. *Critical Public Health*, 21(4), 429–440.

Fischler, C. and Masson, E. (2008). *Manger : français, européens et américains face à l'alimentation*. Odile Jacob, Paris.

Fournier-Schill, M. (2014). Analyse d'une pratique sociale en construction : le cas du tri des déchets dans la famille. PhD Thesis, Lille School of Management Research Center, Lille.

Gram-Hanssen, K. (2010). Residential heat comfort practices: Understanding users. *Building, Research & Information*, 38(2), 175–186.

Hand, M., Shove, E., Southerton, D. (2005). Explaining showering: A discussion of the material, conventional, and temporal dimensions of practice. *Sociological Research Online*, 10(2), 101–113.

Hargreaves, T. (2011). Practice-ing behaviour change: Applying social practice theory to pro-environmental behaviour change. *Journal of Consumer Culture*, 11(1), 79–99.

HCSP (2010). Objectifs de santé publique. Évaluation des objectifs de la loi du 9 août 2004. Propositions, Haut Conseil de la santé publique, Paris.

Hercberg, S. (2006). Eléments de bilan du PNNS (2001–2005) et propositions de nouvelles stratégies pour le PNNS2 (2006–2008) : pour une grande mobilisation nationale de tous les acteurs pour la promotion de la nutrition en France. Report, Ministère de la Santé et des Solidarités, Paris.

IGAS (2010). Evaluation du programme national nutrition santé PNNS2 2006–2010. Report, IGAS RM2010-057P/CGAAER no. 2016.

Keller, M., Kiisel, M., Vihalemm, T. (2015). *From Intervention to Social Change: A Guide to Reshaping Everyday Practices.* Ashgate Publishing, Ltd, Farnham.

Mathe, T., Beldame, D., Hebel, P. (2014). Évolution des représentations sociales du bien manger. *Cahiers de recherche de Crédoc*, C316

Menninger, D.J., Lecoq, G., Guedj, J., Boutet, P., Banel, J.-B. (2010). Evaluation du programme national nutrition santé PNNS2 2006–2010. Report, Inspection générale des Affaires Sociales, Paris.

Ministère du travail, de l'emploi et de la santé (2010). Programme national nutrition santé 2011–2015. Report, Ministère du travail, de l'emploi et de la santé, Paris.

Mugel, O., Gurviez, P., Decrop, A. (2019). Eudaimonia around the kitchen: A hermeneutic approach to understanding food well-being in consumers' lived experiences. *Journal of Public Policy & Marketing*, 38(2), 280–295.

Plessz, M. and Gojard, S. (2014). Fresh is best? Social position, cooking, and vegetable consumption in France. *Sociology*, 49(1), 172–190.

Poulain, J.-P. (2006). Combien de repas par jour ? Normes culturelles et normes médicales en Polynésie Française. *Journal des anthropologues. Association française des anthropologues*, 3–4(106–107), 245–268.

Reckwitz, A. (2002). Toward a theory of social practices: A development in culturalist theorizing. *European Journal of Social Theory*, 5(2), 243–263.

Schatzki, T.R. (1997). Practices and actions: A Wittgensteinian critique of Bourdieu and Giddens. *Philosophy of the Social Sciences*, 27(3), 283–308.

Schatzki, T.R. (2005). Practice mind-ed orders. In *The Practice Turn in Contemporary Theory*, Knorr, K.K., Schatzki, T.R., Von Savigny, E. (eds). Routledge, London.

Shove, E. and Pantzar, M. (2005). Consumers, producers and practices: Understanding the invention and reinvention of Nordic walking. *Journal of Consumer Culture*, 5(1), 43–64.

Shove, E., Pantzar, M., Watson, M. (2012). *The Dynamics of Social Practice: Everyday Life and How it Changes.* Sage Publications, London.

Spurling, N., McMeekin, A., Shove, E., Southerton, D., Welch, D. (2013). Interventions in practice: Re-framing policy approaches to consumer behaviour. Report. Sustainable practices research group, University of Manchester, Manchester.

Turrentine, T.S. and Kurani, K.S. (2007). Car buyers and fuel economy? *Energy Policy*, 35(2), 1213–1223.

Warde, A. (2005). Consumption and theories of practice. *Journal of Consumer Culture*, 5(2), 131–153.

2

"Eating Together" Through the Internet: The Case of Online Weight Loss Support Communities

Steffie GALLIN[1], Laurie BALBO[2] and Marie-Christine LICHTLÉ[3]

[1] *Montpellier Business School, MRM Laboratory, University of Montpellier, France*

[2] *Grenoble École de Management, France*

[3] *MRM Laboratory, University of Montpellier, France*

2.1. Introduction

Worldwide obesity has almost tripled since 1975, with 39% of adults being overweight. In France, 17% of the population is obese which represents more than 8 million people (French Ministry of Health, 2019[1]). In addition, 7 out of 10 women and 1 out of 2 men would like to lose weight (Inserm, 2018[2]). Thus, thanks to the development of Web 2.0, many online communities dedicated to weight loss have appeared.

---

1. See: https://solidarites-sante.gouv.fr/systeme-de-sante-et-medico-social/strategie-nationale-de-sante/priorite-prevention-rester-en-bonne-sante-tout-au-long-de-sa-vie-11031/priorite-prevention-les-mesures-phares-detaillees/article/obesite-prevention-et-prise-en-charge#:~:text=In%20France%2C%2017%20%25%20of%20the%208%20million%20people.

2. See: https://www.inserm.fr/information-en-sante/dossiers-information/obesite.

Evolution of Social Ties Around New Food Practices,
coordinated by Gilles SÉRÉ DE LANAUZE.

Bagozzi and Dholakia (2002, p. 3) define online communities as "social spaces in the digital environment that allow groups to form and be sustained primarily through ongoing communication processes". More specifically, they can be thought of as "gatherings of individuals, enabled by the rise of Web 2.0, who have common interests and goals" (Stenger and Coutant 2013; Gallin *et al.* 2019). These communities have grown greatly thanks to social networks, which allow consumers to interact with a large number of individuals, share their experiences and be inspired by those of others. Very few studies have been devoted to the study of these online communities dedicated to weight loss, despite their strong growth in recent years. Hwang *et al.* (2010) conducted an initial study and were interested in the content exchanged between members of such communities. However, they did not take into account the photos that can be exchanged during discussions and did not study the role of social influence. Our research seeks to fill these gaps.

Thus, the objective of this chapter is to show how users of online communities "eat together" by sharing recipes, photos of the dishes they have made or even tips and advice on nutrition. This sharing contributes to the diversification of their diet. We will focus on online weight loss support communities such as Weight Watchers, forums on Doctissimo or Facebook weight loss support groups, in which food is at the heart of the discussions. We are particularly interested in the phenomenon of social influence that takes place in these communities: how do individuals see their food choices influenced by others? To answer this question, we will rely on excerpts from discussions in communities, as well as on screenshots from weight loss support Facebook groups. Discussions from individual interviews with 23 nutrition experts and 25 users of online weight loss support communities will also be highlighted, underscoring the role of social support in these communities.

2.2. Online weight loss support communities

Health-related online communities are self-help groups in which individuals who have the same illness or are undergoing the same medical treatment come together to seek encouragement, motivation and information, and share experiences (Hwang *et al.* 2010; Huang *et al.* 2015). In this context, several of these communities are organized around the theme of weight loss. They are thus qualified as communities dedicated to weight loss or weight loss support communities.

These communities can be created by dedicated companies or by the users themselves. The first are managed by brands (e.g. Weight Watchers, Dukan, Cohen communities) and very often require a paid subscription: the participants of the slimming programs are encouraged to create an account on the marked community and participate in exchanges by creating discussion threads. Other communities only exist because of the initiative of users and are free: for example, discussion groups on Facebook, doctissimo.fr and aufeminin.com, which can either concern a particular type of diet (such as Weight Watchers, Dukan, Cohen) or weight loss in general. These different communities allow asynchronous conversational exchanges between members who all share the same interest: weight loss. They also offer the possibility, if the Internet user wishes, to remain anonymous (Hwang *et al.* 2010).

According to De Valck *et al.* (2009), virtual communities are very influential reference groups because individuals are free to join the community or not; they are therefore different from *offline* communities that are sometimes imposed (such as the neighborhood, for example). Like any reference group, they are, in a broad sense, "a group of people who significantly influence an individual's behavior" (Bearden and Etzel 1982, p. 184).

2.3. Exchanges in these communities: informational as well as emotional social support

Social support corresponds to material and psychological resources aimed at helping individuals to cope with a stressful event (Cohen 2004). It involves at least two individuals: a sender and a receiver. Support thus refers to people that we can rely on, who let us know that they care about us and appreciate us. The affective dimension is unique to supportive communities according to Shen *et al.* (2010). Social support can be measured quantitatively (number of friends to turn to in times of trouble) and qualitatively (perceived social support) (Zimet *et al.* 1988).

The exchange of social support (informational or emotional) is the main motivation for joining and staying in a virtual community (Ridings and Gefen 2004). The members of these communities exchange advice, opinions and encouragement, which constitute social support. The intensity of these exchanges could be considered as a form of active participation.

In the context of health-related virtual communities, community participation corresponds to the exchange of social support (Kordzadeh *et al.* 2014). In online support communities dedicated to health, this support can be informational in nature (members exchange opinions, advice, tips) and/or emotional in nature (members encourage or comfort each other and show empathy) (Phoenix and Coulson 2008; Kordzadeh *et al.* 2014). Phoenix and Coulson (2008) also distinguish between "esteem support" (messages containing compliments) and "network support" (messages that aim to expand the individual's social network by recommending other members). In this chapter, we will focus on the exchange of informational support and emotional support.

Ballantine and Stephenson (2011) focused on the Weight Watchers Facebook page and conducted a quantitative study with 145 members of the page to learn how social support is conveyed and received. The page provides a communication channel for program members to support each other. They found, through a non-participant observation approach at weekly Weight Watchers meetings over a two-year period, that Weight Watchers support group members seek wellness through weight loss. They distinguish three social support-oriented groups for which there is a strong relationship between the level of participation and social support exchange:

– *Passive recipients* benefit from informational (advice) and emotional (empathy, encouragement) support, but have a role as observers ("lurkers").

– *Active supporters* benefit from informational and emotional support and communicate more with other members.

– *Casual browsers* receive little informational or emotional support and communicate little with other participants.

Ballantine and Stephenson (2011) show that while some users have an active role in the exchange of social support, many members benefit from informational and emotional support without providing support to others themselves (passive role). In this vein and in the context of an online support community for older adults, Pfeil *et al.* (2011) found, among other things, that the role of community members is associated with the type of content those members post. For example, core members (active members who keep the community alive and popular) are more likely to give support, while peripheral members are more likely to post about themselves.

Wright *et al.* (2010) found that participants in support groups prefer support from weak relationships, that is, from acquaintances, distant people, because they bring different perspectives and an objective view. However, Petroczi *et al.* (2006) show that people prefer to seek support from close relationships regardless of the nature of the problem encountered. Tausig and Michello (1988) indicate, however, that both strong and weak relationships may be able to provide informational support.

This link between relationship closeness and social support has been studied by Huang *et al.* (2015), among others. In the Weight Watchers (face-to-face) program, individuals tend to see other participants who are pursuing the same goal as "friends" and seek support from them early in their weight loss. Then, as they progress towards their weight loss goal, uncertainty about achieving that goal decreases and individuals distance themselves from other community members: they no longer share advice and information (informational support).

2.4. Social influence within online weight loss support communities

Social influence exists "when a person adapts their behavior, attitudes, or beliefs to other individuals in a social system" (Trusov *et al.* 2010, p. 646). Thus, an individual's food consumption may be influenced by the presence of others. In this case, social influence exists (McFerran *et al.* 2010).

Little research has examined the role of social influence in online weight loss support communities. According to the research of Wilcox and Stephen (2013), five minutes of Facebook browsing leads people who view content posted online by their close friends to choose less healthy food. It would be interesting to test this result in the case of a weight loss support community and try to understand the social influence that may be present in these communities.

In reality, different processes of social influence can occur during eating behaviors:

– *Social facilitation*: group meals lead to greater food consumption.

– *Impression management*: food consumption is lower when a person eats a meal in the presence of individuals who are watching or not eating.

– *Imitation*: food consumption will adapt to that of a model, both in quantity and quality (choice of balanced foods).

Social influence can be "*informational*" (Burnkrant and Cousineau 1975) through the internalization of advice and counsel that an individual receives (the individual accepts the influence because the behavior is congruent with their value system), as well as "*normative*" when individuals conform to the expectations of others through social identity (Burnkrant and Cousineau 1975). Social identity implies that the *self* becomes interchangeable with that of others (Bagozzi and Dholakia 2002). While informational influence materializes through *internalization*, normative influence results in either *conformity* (the individual accepts the influence in order to get a favorable reaction from others, regardless of personal beliefs) or *identification* with other group members (Burnkrant and Cousineau 1975). Normative influence reflects the need for social recognition (Bearden *et al.* 1989; Shen *et al.* 2010). It corresponds to *social pressure* (Lee *et al.* 2011), that is, being prompted to conform to the expectations of others. Table 2.1 summarizes these different conceptions of social influence.

Park and Lessig (1977) distinguish three types of reference group influence that capture the different types of social influence:

– *The informational influence of the reference group*: the reference group is used as a source of information in the different stages of the decision-making process; the behaviors and opinions of the members of the reference group are then considered as reliable sources of information (informational influence).

– *The utilitarian influence of the reference group*: the reference group establishes the level of norms, values and aspirations of an individual in the process of consumption of products and services (normative or utilitarian influence).

– *The influence linked to the expression of values of the reference group*: the reference group is used to validate, by comparison, the opinions, attitudes, norms and behaviors of the consumers (influence of expression of values); the values of the group and those of the individual are identical.

However, social influence does not have the same effect on each individual. It operates according to different degrees, that is, with varying intensity. Thus, in the literature, we observe that there are *different degrees of susceptibility to social influence*. These have been the subject of a measurement scale created by Bearden *et al.* (1989). However, this scale only applies to the use of products and brands. More recently, Steinberg and Monahan (2007) developed the Resistance to Peer Pressure scale, a measure that is much more general in nature and does not fit any particular type of behavior (examples of items: "Some people think it is more important to be an individual than to fit in"; "Some people will not break the law just because their friends told them to").

Susceptibility to social influence varies across individuals (Bearden *et al.* 1989). However, individual differences in susceptibility to peer influence have been little studied (Prinstein *et al.* 2011). According to Prinstein *et al.* (2011), this susceptibility to social influence is addressed in two ways in the literature: moderators (including peer social status) are examined, and susceptibility to peer influence is studied through the administration of questionnaires. In this way, individuals indicate how they feel influenced by their peers and how they feel they resist this influence (Steinberg and Monahan 2007). Little research has been conducted on how individuals, at different stages, influence each other and how this may vary depending on the strength of the relationship between them (Goode *et al.* 2016).

2.5. A hybrid research methodology

To determine whether food choices can be influenced by members of online weight loss communities, we relied on excerpts from community discussions, screenshots that highlight these exchanges, and individual interviews conducted with 23 nutrition experts and 25 users of online weight loss support communities.

2.5.1. *Step 1: Survey of health experts and community participants*

2.5.1.1. *Data collection*

As part of this data collection, we conducted 23 individual and semi-structured interviews with nutrition experts (dieticians, physicians and professors in university hospitals and public nutrition), and 25 individual and semi-structured interviews with users of online weight loss support communities (Weight Watchers community, Doctissimo, Aufeminin, Carenity, Entrepatients, Linecoaching, Aujourdhui and Le Diet forums). The duration of each interview varied between one and two hours. Most of the interviews were conducted by telephone or Skype due to the geographical dispersion of the respondents.

The "expert" interview guide first included general questions (factors that influence weight gain, digital tools that can help with weight loss), and then questions on the respondents' opinion of these online tools (complementarity with monitoring by a health professional, moderation of these communities).

More specific questions on social influence (online and face-to-face), possible comparisons between individuals in these communities (criteria and standards of comparison) and the effect of participation in these communities on eating behavior were then introduced. The users were contacted through the community platforms. The objective was to better understand how they used these communities and to get their feedback on their weight loss experience. The "user" interview guide, which was less precise than that of the experts, first contained questions on the diets followed, and then questions on the use they make of the community and on the exchanges they have with other members.

2.5.1.2. *Characteristics of the sample*

Fifteen experts were male and eight experts were female. Their ages ranged from 25 to 64 years. These experts included 15 dieticians, four physicians and four professors (Appendix 1, section 2.8.1). Twenty-four of the 25 community users were women and their ages ranged from 21 to 70. In fact, when we looked at the exchanges between users in these communities, it appeared that mainly women animated them. The BMI (Body Mass Index) of the users varied between 20.31 (normal weight) and 42.68 (class III obesity): six people were a normal weight;

13 were overweight; five were class I and II obese and one was class III obese (Appendix 2, section 2.8.2).

2.5.1.3. *The analysis method*

The collected content, which was recorded and fully transcribed, was subjected to two separate content analyses in NVivo 11. We proceeded according to Schilling's (2006) five-level qualitative content analysis spiral to detail the coding procedure (no latent content analysis, interviewees' reactions identified/definition of units of analysis, preliminary categorical system, coding plan and analysis/interpretation). A thematic analysis was chosen to interpret the content and compare opinions. We opted for a plurinominal coding (unit of meaning that can belong to several categories) when necessary.

2.5.2. *Step 2: Analysis of the content shared within the weight loss support communities*

First of all, it is important to underline that the respondents who are members of the communities wish to lose weight for aesthetic and health reasons, but do not suffer from any disorder or disease that justifies a particular diet. The results and recommendations that will be put forward hereafter must therefore be interpreted in light of this information.

To shed light on the discourses of community users and nutrition experts interviewed, we also used screenshots of Facebook posts on weight loss support groups. In order to preserve the anonymity of members, names and profile pictures were blurred out. In addition, the many photos posted by participants, although discussed in this chapter, will not be used for reasons of anonymity.

2.6. Analysis of the results

2.6.1. *Content exchanged in weight loss support communities*

The content shared in these weight loss support communities is diverse, and in particular, people seek advice from other members in a variety of areas, as illustrated in Figure 2.1. Sometimes, it is easier to share about weight loss online as one of the physicians interviewed stated:

> Online, it allows you to broaden your community and your exchanges and discuss more specific or intimate issues.

Figure 2.1. *A search for advice in a weight loss support Facebook group*[3]

The analysis of the interviews conducted with community members shows that different types of advice can be exchanged. The advice can be about diet, weight loss methods or exercise, but can also be about aspects that are not related to weight loss (Table 2.1).

Advice or requests for advice can come from a single person (11 occurrences): "For example, on chrono-nutrition, there is a girl who has been doing chrono-nutrition for months and she knows just about everything inside out, so yes, I am more inclined towards her" (Sandrine, 29 years old). They can also be addressed to a close person (9 occurrences): "To those I know, yes, others I take into account

3. All screenshots are translated versions from the French Facebook support group.

but I think a lot first" (Laura, 31), or to all of the participants (26 occurrences): "To all. I don't favor one girl in particular. I ask a general question. Those who answer me, that's fine. Those who don't answer me, well, that's because they don't have the answer or they don't have the time to answer" (Marina, 32).

Nature of the advice	Number of users (number of user occurrences)	*Verbatims*
Nutritional tips	21 (46)	"For example, one of them advised me not to eat starchy foods at all in the evening" (Victoria, 29 years old).
Advice on the weight loss method	16 (24)	"There are people who advise to go down to, let's say, 1200 kcal per day" (Aurélie, 22 years old).
Exercise tips	8 (12)	"One day I gave advice to a woman to do exercise and this poor lady had not said so, but she was sick, she could not do exercise so it was a bit complicated for her" (Annie, 43 years old).
Advice in the form of exchanges, motivation	7 (12)	"They are there to give you a boost, so that you don't feel guilty, so that you can go back to work tomorrow as if nothing had happened" (Véronique, 40 years old).
Non-weight loss tips	6 (7)	"There is one, her kids were sick. I wrote to her, 'Listen, it would be good if you did this for your children, I also use a lot of essential oils, phytotherapy, homeopathy'" (Pascale, 60 years old).

Table 2.1. *The nature of advice exchanged in online communities (Gallin* et al. *2019)*

The posts on weight loss support Facebook groups are numerous and above all varied. They are systematically commented on by other members of the community. The posts range from a simple "summary" that consists of indicating results over a given period of time, to an exchange of support (informational and/or emotional) and the sharing of recipes that contribute to the diversification of our diet. Figures 2.2 and 2.3 correspond, respectively, to a "summary" related to weight loss and a "summary" related to the day's diet (nature, quantities and calories).

Rising star · 31 May, 11:27 · Support and discussion

Hi girls,
It's good to see -500g on the scales.
I do 45 mins of cardio (30 mins from a New Zealand channel and a 15 minute bikini workout with sis). Then I do 15 mins of stretching with TiboInShape.
20/03/2020-88.6kg
20/04/2020-84.9kg
20/05/2020-82.00kg
31/05/2020-81.5kg
GOAL 54/60kg SUMMER 2021
Have a good day

Figure 2.2. *A "summary" related to weight loss in a weight loss support Facebook group*

22 June, 20:51

A day on my plate

BREAKFAST:

- 1 corn cake + 42g of goat's cheese + 6g of liquid honey
- 1 corn cake + 1 tea of homemade cherry jam + some chocolate chips + 1 pinch of grated coconut

LUNCH:

- 355g of curried vegetables and coconut milk (potato + pepper + tomato + courgette {from the garden})
- 2 slices of roast pork
- 1 Greek natural yoghurt
- 103g of cherries

DINNER:

- 1 glass of citrus soda
- 379g of curried vegetables and coconut milk (leftover from lunch)
- 2 barbecued beef kebabs (leftover from yesterday's lunch)
- 1 corn cake + 33g of goat's cheese + 1 tea of homemade cherry jam

I finished my day on 1415 calories
I've reset my application to 0 because since Thursday evening I've despaired about weighing everything and didn't pay attention
but I'm back at it today
Have a good evening everyone and tomorrow is a new day on my plate

Figure 2.3. *A diet-related "summary" in a weight loss support Facebook group. For a color version of this figure, see www.iste.co.uk/seredelanauze/evolution.zip*

2.6.2. *The exchange of informational support in online weight loss support communities*

Many members of these communities seek informational support, as shown in Figure 2.4. The individual is looking for "solutions" and in particular, to "limit cravings". They therefore ask for tips, which correspond to a search for informational support.

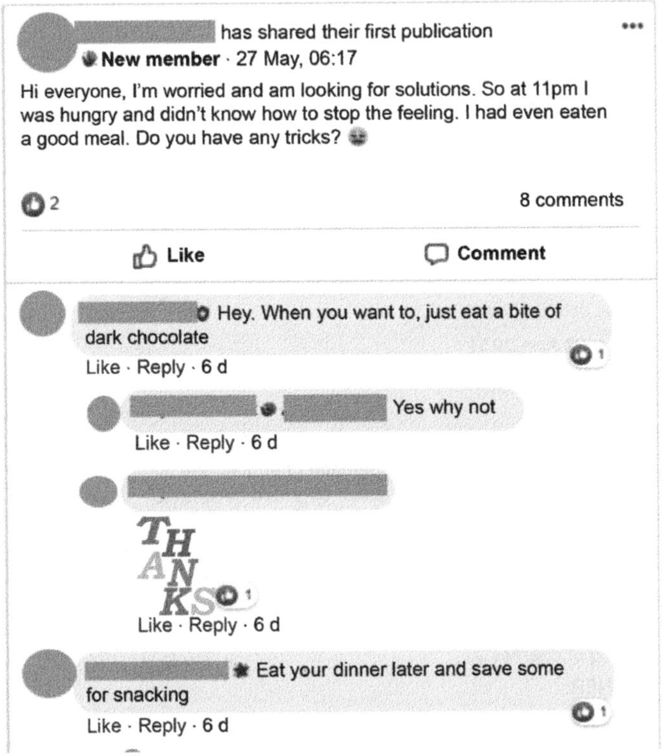

Figure 2.4. *Seeking informational support in a weight loss support Facebook group: the "tips". For a color version of this figure, see www.iste.co.uk/seredelanauze/evolution.zip*

Multiple publications related to the search for informational support appear every day. They are a motivation to participate in this type of community. Individuals want to exchange and get other members' opinions on their weight loss method, as shown in these other examples of information support research (Figures 2.5 and 2.6).

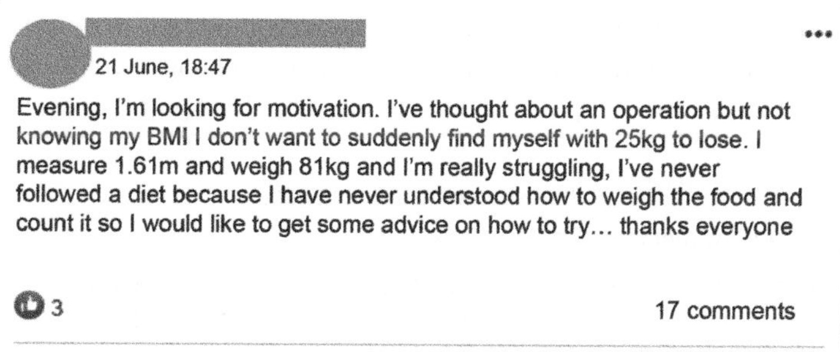
21 June, 18:47

Evening, I'm looking for motivation. I've thought about an operation but not knowing my BMI I don't want to suddenly find myself with 25kg to lose. I measure 1.61m and weigh 81kg and I'm really struggling, I've never followed a diet because I have never understood how to weigh the food and count it so I would like to get some advice on how to try... thanks everyone

3 17 comments

Figure 2.5. *Another example of seeking informational support in a weight loss support Facebook group: seeking various tips*

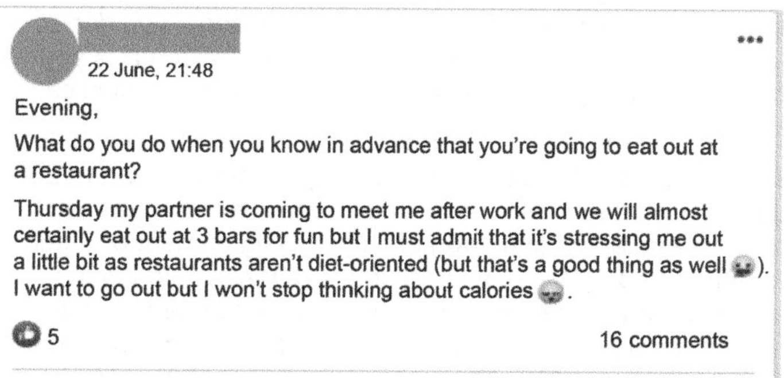
22 June, 21:48

Evening,

What do you do when you know in advance that you're going to eat out at a restaurant?

Thursday my partner is coming to meet me after work and we will almost certainly eat out at 3 bars for fun but I must admit that it's stressing me out a little bit as restaurants aren't diet-oriented (but that's a good thing as well). I want to go out but I won't stop thinking about calories .

5 16 comments

Figure 2.6. *Another example of seeking informational support in a weight loss support Facebook group: seeking opinions/experiences when dining out*

Individuals who are members of weight loss support groups can seek informational support as well as give it, without necessarily requesting informational support first. In Figure 2.7, a group member published their "food rebalancing program" with details of their diet (nature and quantities consumed), as well as what they consider good practices to promote weight loss (e.g. "at each meal, you need 2 teaspoons of oil").

The other example below also shows an informational support input not previously requested: the name of a mobile application for counting steps, Pacer. It is therefore a piece of advice related to the sport activity since the participant advises the use of a specific application. We have also identified this type of advice in Table 2.1, from the interviews with users of this type of community.

While there are many examples of research and informational support, participants in these types of communities also exchange emotional support. Sometimes, informational support can lead to emotional support.

has shared their first publication ...

New member · 28 May, 12:54

Hi everyone, I'm posting my diet program which I followed to lose 10kg. I had posted it on a publication and many people were interested so here it is, I hope that it can help other people

Morning: hot drink of choice (coffee or 150ml of milk or tea with 1 sugar max.) 80g of bread of choice (except wholemeal bread) or 3 biscuits with 10g of butter or 30g of cheese or 3tsp of jam + 150g of fruit or 150ml of 100% pure fruit juice and if you don't eat cheese or drink milk in the morning you can also eat a yoghurt. If you eat a fruit yoghurt of course cut out the fruit or fruit juice.

Midday: 50g of raw starch (rice, pasta, semolina, ebly) or 200g of mashed potatoes or 150g of cut potatoes or 60g of bread or 250g of legumes (corn, flageolet beans, white beans, lentils, peas, etc) Choice of vegetables and 140g of fish or 120g of meat or 3 eggs. 150g of fruit and 1 item of dairy produce (cheese if you didn't eat it in the morning (because you can only have it once a day) if not, any natural yoghurt but it must be 0% (that for a slim figure and fresh cheese at 0%). This is the same idea as the morning, if you eat a fruit yoghurt forego the fruit.

Evening: 120g of fish or 100g of meat and choice of vegetables + 0% dairy product (avoid fruit yoghurts in the evening as they are sugary)

For each meal you must cook with 2tsp of oil (preference for olive oil...)

Avocado is allowed at ½ in the afternoon 2x per week because it is calorific

These vegetables are not allowed in the evening because they are too sugary or calorific:

Carrots
Squash
Heart of palm
Salsify
Beets

Figure 2.7. *The provision of spontaneous informational support in a weight loss support Facebook group: details of a dietary program*

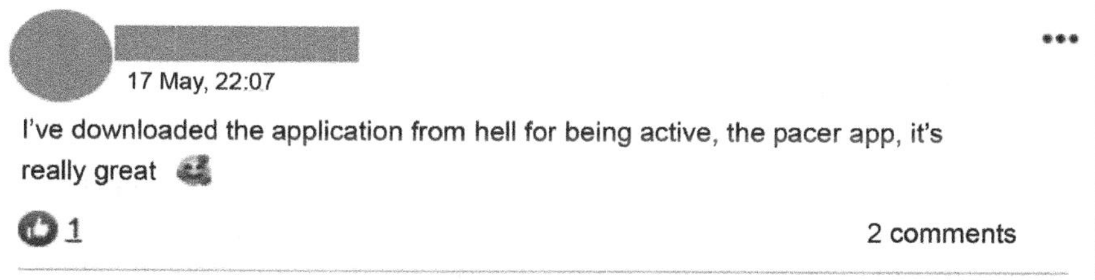

Figure 2.8. *The provision of spontaneous sport-related informational support in a weight loss support Facebook group: advice to use the Pacer app*

2.6.3. *The exchange of emotional support in online weight loss support communities*

These two types of support can become interdependent: "I think it's a way, by advising others, of motivating yourself" (Edith, 53). These two types of support were highlighted by Kordzadeh *et al.* (2014), who showed that participation in an online community dedicated to health corresponded to the exchange of both informational (sharing knowledge and experiences) and emotional (empathy, affection, encouragement) social support. Hwang *et al.* (2010), in the context of an online weight loss support community (SparkPeople.com), also highlighted these exchanges of encouragement and motivation between participants.

In the interviews with community users, the main motivation for this participation that appears in the users' discourse is emotional support (23 respondents, 64 occurrences), illustrated in particular by the following verbatim: "I think that even if we are not all the same weight or the same size, and do not have the same objectives, just sharing our difficulties does something. It can lift your spirits. You can get motivated again. I think the fact that it's more than a blog and that there is interaction between the people in the community is very important" (Yasmine, 35). Participants in these communities seek to "encourage each other" (Jacques, 59). One of the nutrition experts interviewed indicates that "It [the communities] can allow people to connect with each other and increase motivation. The group effect can be more motivating than if they're all alone at home" (Dietitian).

In the following publication (Figure 2.9) on a weight loss support group, the participant shares her progress in losing weight and in sports activities. We can therefore deduce that she is seeking emotional support through encouragement to continue her efforts. Indeed, she receives comments that correspond to an emotional support with congratulations and encouragement.

Figure 2.9. *Seeking and providing emotional support in a weight loss support Facebook group: sharing weight loss progress. For a color version of this figure, see www.iste.co.uk/seredelanauze/evolution.zip*

In Figure 2.10 (as of January 6, 2020), the person is seeking emotional support because she is unable to lose weight. In the comments that were subsequently posted, some group members provide emotional support by urging her not to get discouraged.

The exchange of support can also be both informational and emotional (see Figure 2.11).

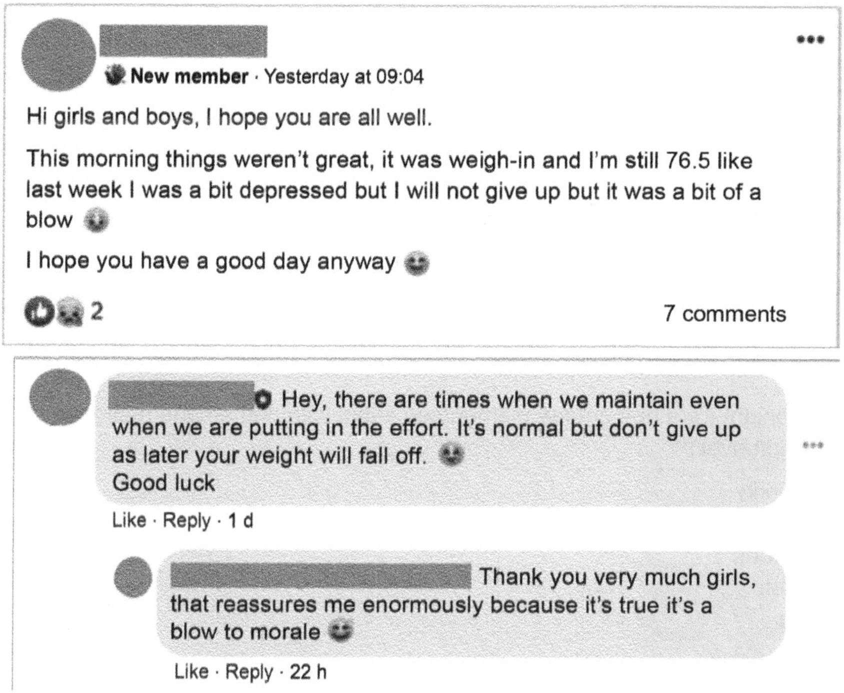

Figure 2.10. *Seeking emotional support in a weight loss support Facebook group: being encouraged even when not losing weight*

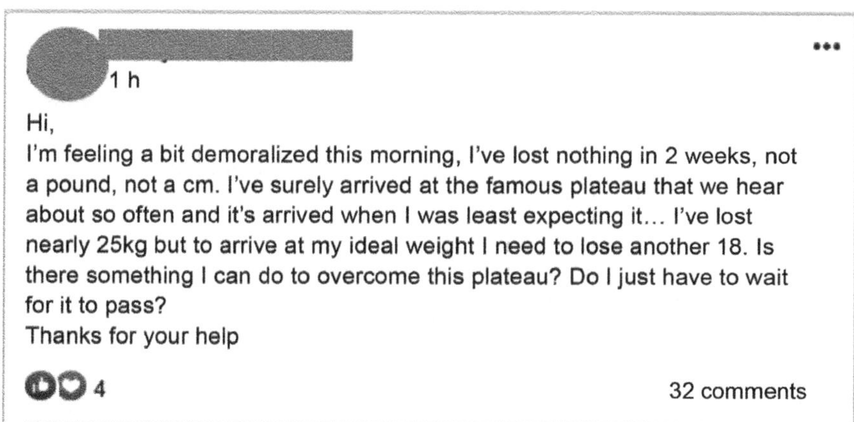

Figure 2.11. *Seeking both informational and emotional support in a weight loss support Facebook group*

2.6.4. *Recipes at the heart of discussions in the communities…*

In these groups, individuals support each other, but they also share recipes. For eight users, the communities allow them to diversify their diet by sharing recipes: "We talk about recipes a lot" (Jacqueline, 58). These recipes can be approximate (indication of ingredients only; see Figure 2.12) or precise (indication of ingredients, quantities and method; see example 2.13).

Figure 2.12. *Sharing a rough recipe in a weight loss support Facebook group*

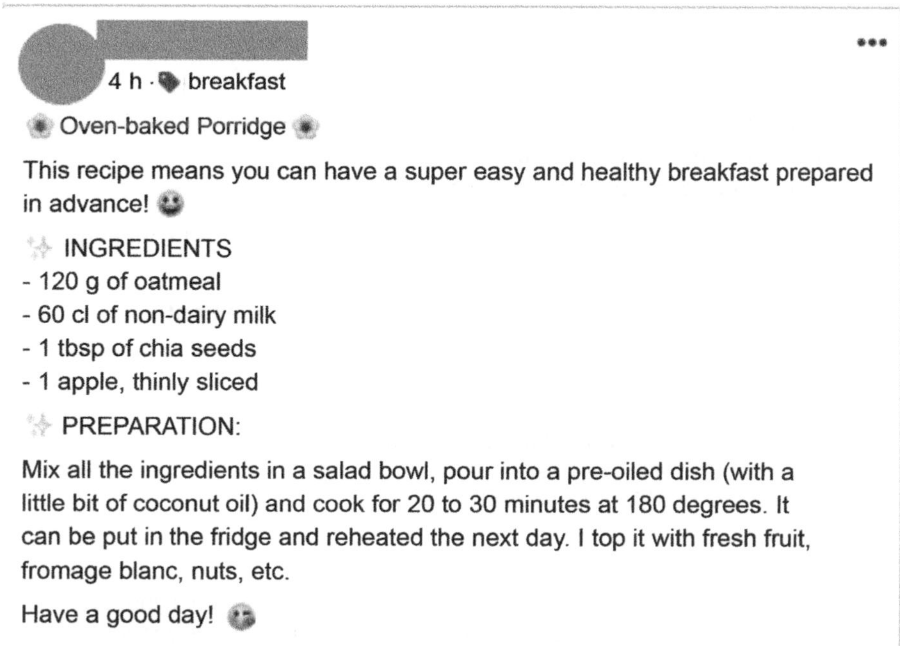

4 h · breakfast

Oven-baked Porridge

This recipe means you can have a super easy and healthy breakfast prepared in advance!

INGREDIENTS
- 120 g of oatmeal
- 60 cl of non-dairy milk
- 1 tbsp of chia seeds
- 1 apple, thinly sliced

PREPARATION:

Mix all the ingredients in a salad bowl, pour into a pre-oiled dish (with a little bit of coconut oil) and cook for 20 to 30 minutes at 180 degrees. It can be put in the fridge and reheated the next day. I top it with fresh fruit, fromage blanc, nuts, etc.

Have a good day!

Figure 2.13. *Sharing a specific recipe in a weight loss support Facebook group. For a color version of this figure, see www.iste.co.uk/seredelanauze/evolution.zip*

Some members also ask for recipes, but may go beyond that by asking other members about their eating or shopping habits. Figure 2.14 highlights a member who asks for a recipe as well as where others usually buy ingredients. Participating in this type of community is not just blogging (many food blogs are available online); it is an exchange about eating habits, shopping habits and daily life.

This sharing encourages food variety and is encouraged by nutrition experts: "Afterwards, what can be quite good is all of the recipe blogs and others that can help them to diversify the diet, without necessarily being diet recipe blogs, but there you go, to open up to other horizons" (Dietitian). A dietitian would advise these communities "As long as there is no advice behind it and then a new method that blooms on every page. If it's really just a support between patients and then yes, that's it. Or recipes, things like that, I think it's a good thing." The communities allow for the acquisition of nutritional knowledge: "It conveys the concept so we can learn things" (Dietitian). One of the teachers interviewed encourages communities "of people who are trying to get better in their relationship with food".

Figure 2.14. *A request for a yogurt recipe in a weight loss support Facebook group*

While the exchange of opinions and advice refers to informational influence, we will also see that community users can be influenced through norms that they have internalized (normative influence).

2.6.5. *Informational and normative social influence in online weight loss support communities*

Normative influence is the tendency to conform to the expectations of others and reflects the psychological need to conform to group standards and norms in order to be recognized (Bearden *et al.* 1989; Shen *et al.* 2010). Participants do not seek social recognition (16 occurrences), as 20 respondents indicate that it does not matter to them if other participants do not agree with their dietary choices or the chosen method of weight loss. However, the following discussion on a weight loss Facebook group suggests otherwise (Figure 2.15).

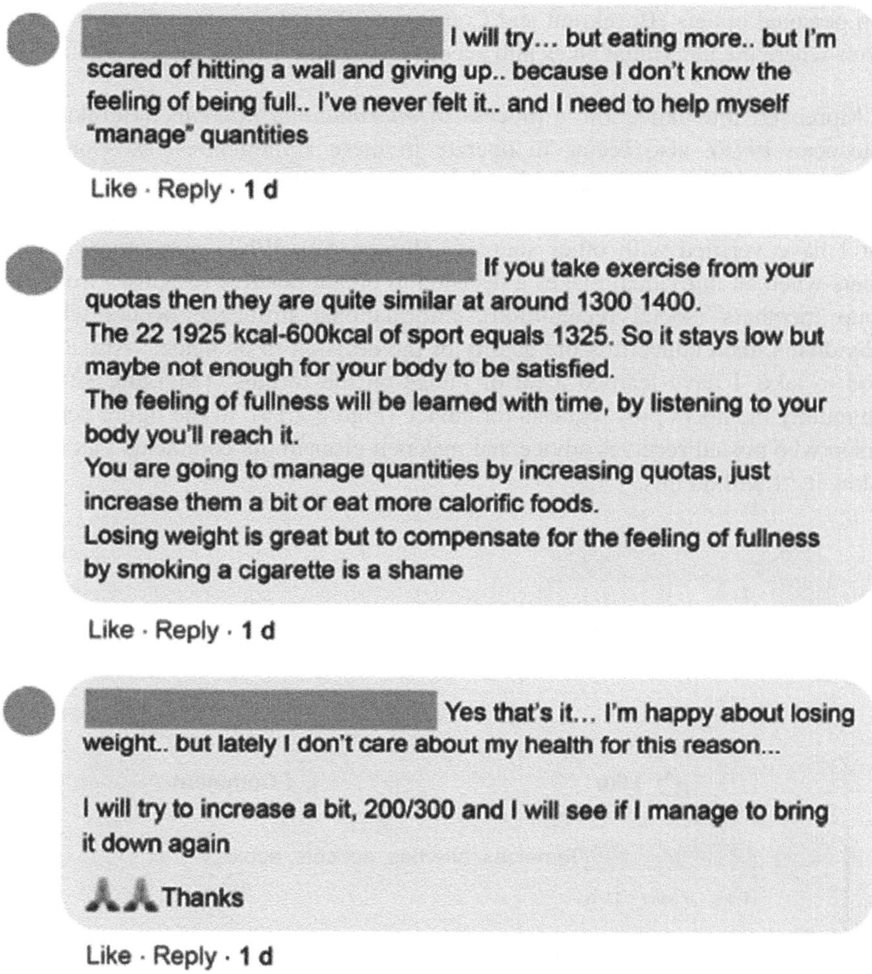

Figure 2.15. *An example of normative influence in a weight loss support Facebook group*

In this group, the participants all follow the same weight loss method, which consists of never exceeding 1,800 kcal per day and never going below 110 kcal. The individuals follow this method and its rules, the principles of which are regularly reminded by the group moderator. Beyond the calories consumed, food variety and "homemade" are widely advocated; members comply with these rules by publishing their quotas or even photos of their plates that are in line with these rules. Compliance reflects the normative influence of other members, regardless of our own personal beliefs (Burnkrant and Cousineau 1975). It corresponds to subjective norms where the individual takes into account what others think they should do.

Moreover, internalization, a process of *informational influence* (Burnkrant and Cousineau 1975), also seems to operate in these communities, as soon as the participants consider the information delivered as reliable and sure. The following verbatim illustrates this point: "I follow the advice I receive that seems reasonable or that I have verified with other sources" (James, 59). While normative influence occurs when an individual makes a decision to obtain positive responses from other group members (social recognition), informational influence occurs when the individual's main concern is the quality of the decision to be made: "Any advice is good to take, I have learned a lot of things on the forums" (Myriam, 32). Many community members post requests for advice (Figure 2.16). In the figure below, the person who posted receives advice and makes it clear in the comments that she will follow it: "I will do this."

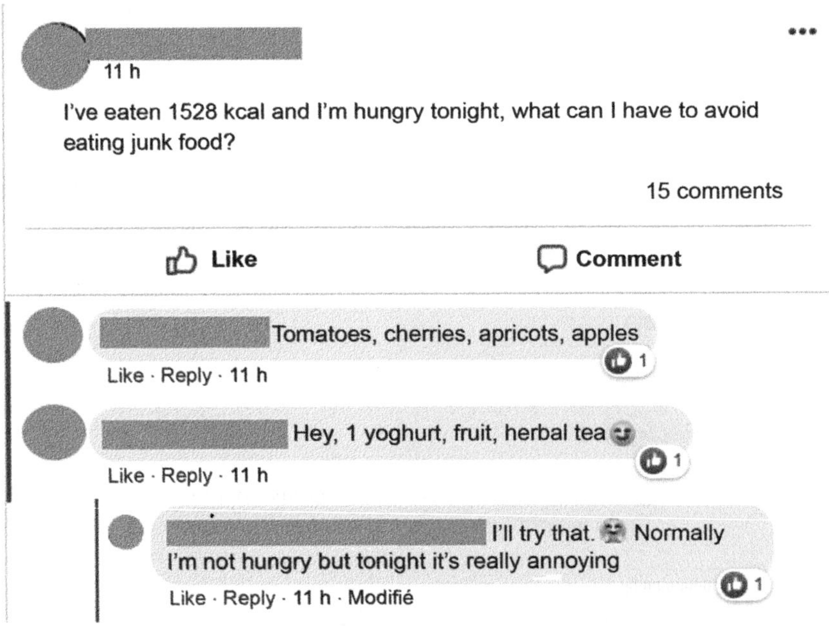

Figure 2.16. *An example of informational influence in a weight loss support Facebook group*

Twenty-one respondents receive and give advice about the community. Four respondents who have not received any advice indicate that they do not give advice either: "I don't normally give advice, or I really don't give advice when there are menus that are completely out of whack" (Denise, 62). Still, we can think that participants usually exchange advice (giving and receiving) in this type of community, thus reinforcing the phenomenon of informational social influence.

Some participants are perceived as models and thus become pseudo-experts, opinion leaders once they have lost a certain amount of weight: "Especially in the sense that I have lost nearly 30 kilos and so people asked me how I had done it, had I been on a diet, and so I gave them a little of what I did on a daily basis to help them as best as possible" (Elsa, 21); "In fact, when people see that you have lost a certain number of kilos, they come to you by themselves. When it is written that I have lost 33 kilos, it is sure that for them I am an example" (Véronique, 40). Figure 2.17 refers to a person who shares her evolution in terms of weight loss (more than 10 kg lost). Some of the comments related to this publication actually show requests for advice.

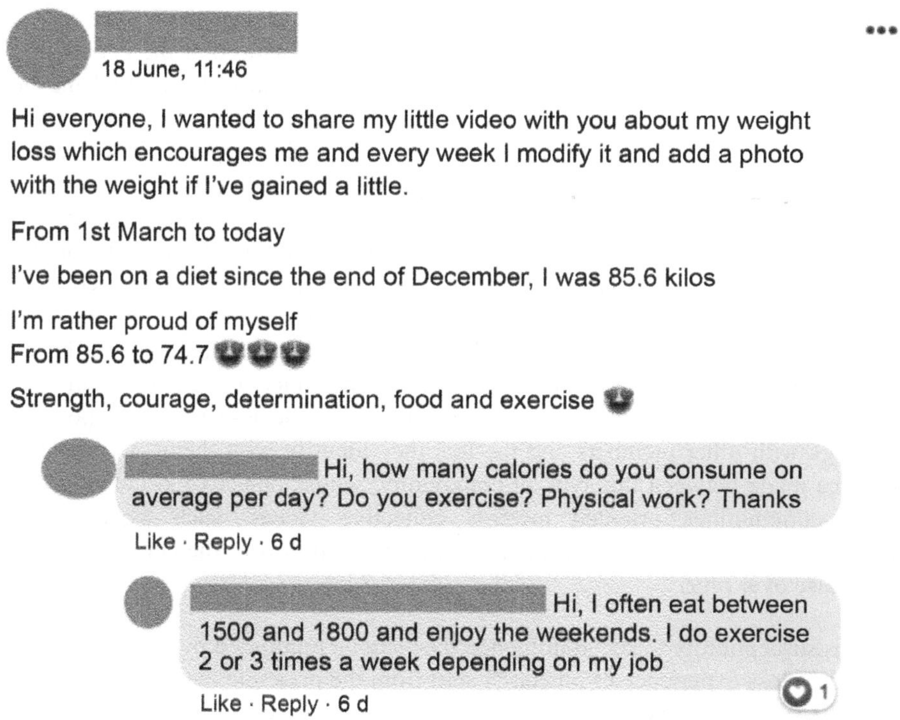
18 June, 11:46

Hi everyone, I wanted to share my little video with you about my weight loss which encourages me and every week I modify it and add a photo with the weight if I've gained a little.

From 1st March to today

I've been on a diet since the end of December, I was 85.6 kilos

I'm rather proud of myself
From 85.6 to 74.7

Strength, courage, determination, food and exercise

Hi, how many calories do you consume on average per day? Do you exercise? Physical work? Thanks

Like · Reply · 6 d

Hi, I often eat between 1500 and 1800 and enjoy the weekends. I do exercise 2 or 3 times a week depending on my job

Like · Reply · 6 d

1

Figure 2.17. *A publication that triggers a request for information*

However, are all users equally influenced in their food choices? Are they also influencers?

2.6.6. *The degree of susceptibility to social influence*

Twenty-three people believe that participants' eating behaviors in the community can be influenced by other members. Sixteen respondents believe that they and others can be influenced in their food choices, and only two people indicated that the forums do not influence them or others. However, these two respondents who indicated that the forums did not impact their diets were not actively participating in the community. Although passive participants, who simply read posted messages, may be influenced, we assume that they are influenced to a much lesser extent than active participants who post content, support each other and identify with other members.

Susceptibility to the influence of the community could depend on the *degree of progress towards the weight loss goal*: are we more influenced at the beginning of the weight loss process when we glean a lot of advice in order to find the best solution for losing weight? Are we less influenced when we have lost a large part of our desired weight knowing that we have acquired solid knowledge? Twelve users consider themselves to be at an advanced stage in their weight loss and are therefore less susceptible to influence now than at the beginning of their weight loss: "Personally, no, because I've been trying to get my head around it for a while now, but I think that if I had gone to the forum a few years ago, I could have been much more influenced than that" (Marion, 22); "I don't think I've been doing that much anymore. Since the time I practiced" (Pascale, 60). These results confirm those of Huang *et al.* (2015) who show, in the Weight Watchers face-to-face community, that the closer individuals get to their weight loss goal, the less social support they exchange with other members and the less they identify with them. A comparison could be made between this type of community, whose objective is individual and online communities oriented towards collective values, such as those that vegetarians exchange on. In the latter case, identification with the community tends to increase over time.

This susceptibility to social influence seems to be reinforced by the perception of the weight loss experience of others: "I start from the principle that you have to trust people, I trust them. So for someone who has lived an experience and who talks about it on doctissimo, it's real" (Sylvie, 56), as well as by the satisfaction of the advice received (nine respondents): "In general, I am better informed than by the nutritionists. I regret to say so" (Jacques, 59). Figure 2.18. corresponds to an active member who thanks the group for its support: "Thank you to this very kind group which has given me wings!"

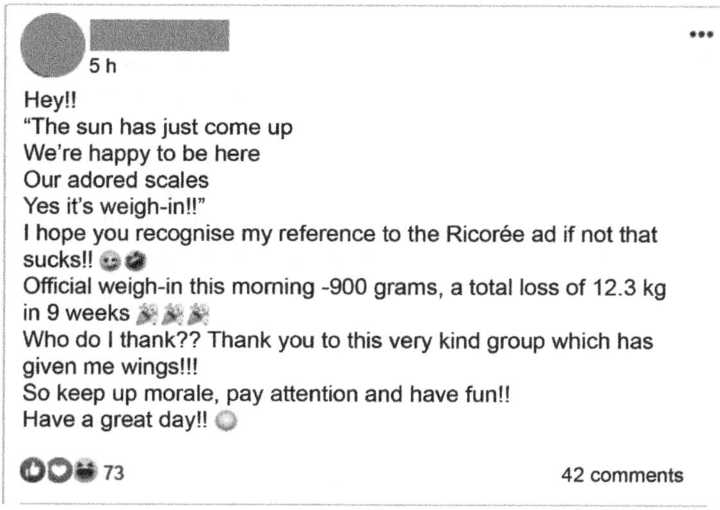

Figure 2.18. *A member's thank you for the support received on the group. For a color version of this figure, see www.iste.co.uk/seredelanauze/evolution.zip*

Even the two respondents who are not actively involved are looking for the exchanges that take place at these gatherings: "It always helps to see other situations. Just seeing people talking and seeing how they are trying to cope helps a bit" (Mélanie, 40). This notion of trust evoked by users is an essential component in the propensity to give and receive information in a virtual community (Foster *et al.* 2010).

2.7. Conclusion

This chapter shows how individuals can eat better "together" through sharing in online weight loss support communities. Although virtual, individuals eat better "together" through the sharing of their meals and eating plans with the community. Through interviews with users of this type of community and nutrition experts, as well as publications from weight loss support Facebook groups, we were able to highlight the exchange of support, informational (advice, tips) and/or emotional (encouragement), which contributes to the modification of eating behaviors. Support is the main motivation for participating in this type of community.

Dietary diversification is widely advocated with numerous recipe sharing. Some members even go so far as to publish a photo of their plate every day, for each meal (breakfast, lunch and dinner), indicating the recipe in the comments. These groups are not only places to exchange information about food and weight loss, but also places to live, since everything can be shared spontaneously. Individuals can therefore eat together, online, by sharing their food at a given moment.

These places are thus the scene of social influence, both informational and emotional. The degree of progress towards weight loss determines the degree of susceptibility to social influence, and the more important the weight loss is, the more the members will be solicited to share the keys to their success. They will therefore eat even more, virtually with the other members of the weight loss support community.

2.8. Appendices

2.8.1. *Appendix 1: Characteristics of the expert sample*

Occupation of expert respondents	Age	Gender	Place of practice
Professor (university hospital) 1	56 years old	M	Rouen (76)
Professor (university hospital) 2	52 years old	M	Montpellier (34)
Professor (university hospital) 3	44 years old	M	Rennes (35)
Professor 4	64 years old	M	Paris (75)
Doctor (private practice) 1	56 years old	M	Paris (75)
Physician (university hospital) 2	29 years old	F	Montpellier (34)
Physician (university hospital) 3	27 years old	F	Montpellier (34)
Physician (university hospital) 4	33 years old	M	Montpellier (34)

Occupation of expert respondents	Age	Gender	Place of practice
Dietitian 1	41 years old	M	Lille (59)
Dietitian 2	37 years old	F	Tarnos (40)
Dietician 3	26 years old	F	Dijon (21)
Dietician 4	25 years old	M	Grenoble (38)
Dietician 5	37 years old	M	Tours (37)
Dietician 6	35 years old	M	Toulouse (31)
Dietician 7	37 years old	M	La Rochelle (17)
Dietician 8	35 years old	M	Paris (75)
Dietician 9	44 years old	M	Grenoble (38)
Dietician 10	52 years old	F	Montpellier (34)
Dietician 11	38 years old	F	Paris (75)
Dietician 12	37 years old	M	Poitiers (86)
Dietician 13	29 years old	F	Dijon (21)
Dietician 14	30 years old	M	Lyon (69)
Dietician 15	30 years old	F	Poitiers (86)

2.8.2. *Appendix 2: Characteristics of the user sample*

First name	Age	Occupation	BMI	Community
Julie	21 years old	Nursery assistant	38.58	Doctissimo.fr
Claude	70 years old	Retired	29.07	Carenity.fr
Véronique	40 years old	Housewife	29.85	Aujourdhui.com
Yasmine	35 years old	Engineer	26.3	Aujourdhui.com
Melanie	40 years old	Housewife	27.36	Carenity.fr
Pascale	60 years old	Disabled	42.68	Doctissimo.fr and LeDiet.fr
Muriel	41 years old	Rental manager	27.14	Doctissimo.fr
Myriam	32 years old	Self-employed entrepreneur	22.49	Doctissimo.fr
Jacqueline	58 years old	Retired	28.98	Weight Watchers.fr
Aurélie	22 years old	Job seeker	28.98	Doctissimo.fr
Elsa	21 years old	Job seeker	20.31	Doctissimo.fr
Laura	31 years old	Optician	25	Doctissimo.fr
Jacques	59 years old	Teacher	23.3	Entrepatients.fr
Edith	53 years old	Job seeker	34.29	Weight Watchers.fr
Annie	43 years old	Employee in a daycare center	24.45	LeDiet.fr
Victoria	29 years old	Housewife	26.47	Aufeminin.com
Sylvie	56 years old	Public servant	28.84	Doctissimo.fr

First name	Age	Occupation	BMI	Community
Aude	22 years old	Quantity surveyor	33.9	Aufeminin.com
Marion	22 years old	Sales consultant	29.73	Aufeminin.com
Nadège	32 years old	Chef	23.61	Aufeminin.com
Marina	32 years old	On maternity leave	30.86	Aufeminin.com
Sandrine	29 years old	Caregiver	25.46	Aufeminin.com
Denise	62 years old	Retired	21.88	Weight Watchers.fr
Emilie	23 years old	Student	26.93	Aufeminin.com
Valérie	56 years old	Clerk in a notary's office	31.25	Weight Watchers.fr and Linecoaching.com

2.9. References

Bagozzi, R.P. and Dholakia, U.M. (2002). Intentional social action in virtual communities. *Journal of Interactive Marketing*, 16(2), 2–21.

Ballantine, P.W. and Stephenson, R.J. (2011). Help me, I'm fat! Social support in online weight loss networks. *Journal of Consumer Behaviour*, 10, 332–337.

Bearden, W.O. and Etzel, M.J. (1982). Reference group influence on product and brand purchase decisions. *Journal of Consumer Research*, 9(2), 183–194.

Bearden, W.O., Netemeyer, R.G., Teel, J.E. (1989). Measurement of consumer susceptibility to interpersonal influence. *Journal of Consumer Research*, 15(4), 473–481.

Burnkrant, R.E. and Cousineau, A. (1975). Informational and normative social influence in buyer behavior. *Journal of Consumer Research*, 2(3), 206–215.

Cohen, S. (2004). Social relationships and health. *American Psychologist*, 59(8), 676–684.

De Valck, K., Van Bruggen, G.H., Wierenga, B. (2009). Virtual communities: A marketing perspective. *Decision Support System*, 47, 185–203.

Foster, M.K., Francescucci, A., West, B.C. (2010). Why users participate in online social networks. *International Journal of e-Business Management*, 4(1), 3–19.

Gallin, S., Lichtlé, M.-C., Balbo, L. (2019). Le rôle de la comparaison sociale dans les communautés en ligne de soutien à la perte de poids. *Décisions Marketing*, (95), 33–45.

Goode, M.R., Hart, K., Thomson, M. (2016). Say more! The liability of strong ties on desire for special experiences. *Journal of Consumer Psychology*, 26(1), 91–97.

Huang, S.C., Broniarczyk, S.M., Zhang, Y., Beruchashvili, M. (2015). From close to distant: The dynamics of interpersonal relationships in shared goal pursuit. *Journal of Consumer Research*, 41(5), 1252–1266.

Hwang, K.O., Ottenbacher, A.J., Green, A.P., Cannon-Diehl, M.R., Richardson, O., Bernstam, E.V., Thomas, E.J. (2010). Social support in an Internet weight loss community. *International Journal of Medical Informatics*, 79, 5–13.

Kordzadeh, N., Liu, C.Z., Au, Y.A., Clark, J.G. (2014). A multilevel investigation of participation within virtual health communities. *Communications of the Association for Information Systems*, 34(26), 493–512.

Lee, D., Kim, H.S., Kim, J.K. (2011). The impact of online brand community type on consumer's community engagement behaviors: Consumer-created vs. marketer-created online brand community in online social-networking web sites. *CyberPsychology, Behavior, and Social Networking*, 14(1–2), 59–63.

McFerran, B., Dahl, D.W., Fitzsimons, G.J., Morales, A.C. (2010). I'll have what she's having: Effects of social influence and body type on the food choices of others. *Journal of Consumer Research*, 36(6), 915–929.

Park, C.W. and Lessig, V.P. (1977). Students and housewives: Differences in susceptibility to reference group influence. *Journal of Consumer Research*, 4(2), 102–110.

Petroczi, A., Nepusz, T., Bazso, F. (2006). Measuring tie-strength in virtual social networks. *Connections*, 27(2), 39–52.

Pfeil, U., Svangstu, K., Ang, C.S., Zaphiris, P. (2011). Social roles in an online support community for older people. *International Journal of Human-Computer Interaction*, 27(4), 323–347.

Phoenix, K.H.M.O. and Coulson, N.S. (2008). Exploring the communication of social support within virtual communities: A content analysis of messages posted to an online HIV/AIDS support group. *CyberPsychology and Behavior*, 11(3), 371–374.

Prinstein, M.J., Brechwald, W.A., Cohen, G.L. (2011). Susceptibility to peer influence: Using a performance-based measure to identify adolescent males at heightened risk for deviant peer socialization. *Developmental Psychology*, 47(4), 1167–1172.

Ridings, C.M. and Gefen, D. (2004). Virtual community attraction: Why people hang out online. *Journal of Computer Mediated Communication*, 10(1).

Schilling, J. (2006). On the pragmatics of qualitative assessment. Designing the process for content analysis. *European Journal of Psychological Assessment*, 22(1), 28–37.

Shen, K.N., Yu, A.Y., Khalifa, M. (2010). Knowledge contribution in virtual communities: Accounting for multiple dimensions of social presence through social identity. *Behaviour and Information Technology*, 29(4), 337–348.

Steinberg, L. and Monahan, KC. (2007). Age differences in resistance to peer influence. *Developmental Psychology*, 43(6), 1531–1543.

Stenger, T. and Coutant, A. (2013). Médias sociaux : clarification et cartographie. Pour une approche sociotechnique. *Décisions Marketing*, 70, 107–117.

Tausig, M. and Michello, J. (1988). Seeking social support. *Basic and Applied Social Psychology*, 9(1), 1–12.

Trusov, M., Bodapati, A.V., Buckin, R.E. (2010). Determining influential users in Internet social networks. *Journal of Marketing Research*, 47(4), 643–658.

Wilcox, K. and Stephen, A.T. (2013). Are close friends the enemy? Online social networks, self-esteem, and self-control. *Journal of Consumer Research*, 40, 90–103.

Wright, K.B., Rains, S., Banas, J. (2010). Weak-tie support network preference and perceived life stress among participants in health-related, computer-mediated support groups. *Journal of Computer-Mediated Communication*, 15(4), 606–624.

Zimet, G.D., Dahlem, N.W., Zimet, S.G., Farley, G.K. (1988). The multidimensional scale of perceived social support. *Journal of Personality Assessment*, 52(1), 30–41.

3

"Eating Together": With or Without the Dietary Constraints of Others?

Andréa GOURMELEN[1], Marie-Christine LICHTLÉ[1], Laurie BALBO[2] and Steffie GALLIN[3]

[1] *MRM Laboratory, University of Montpellier, France*

[2] *Grenoble École de Management, France*

[3] *Montpellier Business School, MRM Laboratory, University of Montpellier, France*

"To eat or not to eat, the important thing is to be together"
Indonesian proverb

3.1. Introduction

A proportion of 20–30% of French people have a sugar intake of more than 100 grams per day, whereas 50 grams would be sufficient in most cases. Therefore, diets that allow for reducing sugar consumption have appeared. This is the case, for example, of the "low-carb" diet (or Paleolithic diet), which is inspired by the diet of our ancestors 2.5 million years ago. Bread, cakes, pasta, rice, potatoes and some fruits are forbidden. Beyond simple weight loss, this diet has become a real way of life for some people and has completely changed their food choices.

Evolution of Social Ties Around New Food Practices,
coordinated by Gilles SÉRÉ DE LANAUZE.

This type of dietary constraint can be chosen when the individual imposes a diet on themselves out of personal conviction. If we take the example of the low-sugar diet, an individual may opt for it because they fear for their future health (e.g. fear of weight gain, or certain diseases such as diabetes and cancer). Conversely, an individual may adopt this same diet not through choice but through medical obligation, if they have health problems (such as diabetes). They then endure the dietary constraints associated with this diet.

This reasoning (constraints endured or chosen) can be applied to foodstuffs other than sugar and thus, today, people can build their own diet with more or less restriction, endured because of health problems or chosen on the basis of personal conviction (food judged bad for health, for the environment, etc.). This results in a change of "eating behavior".

Eating behavior represents all the behaviors of an individual with respect to their food consumption. Its main function is to ensure the necessary energy supply for the body. It is defined in a narrow sense as "a discontinuous series of consumption episodes of variable duration, separated from other consumption episodes by intervals of equally variable duration" and in a broad sense as "the behaviors associated with the search for and acquisition of food" (Bellisle and Dalix 2006, p. 159).

However, despite various dietary constraints, the eating habits of the French are dictated by aspects other than purely nutritional ones. Indeed, the pleasures of the table are valued by the French, whether it is to take a meal at the canteen, at home or in a restaurant. An OECD study[1] shows that the French are the champions in terms of time spent at the table each day: they spend an average of 2 hours a day eating and drinking. However, another study conducted by Nielsen[2] in 2019 shows that while 44% of households have eating habits that can be described as "classic", 56% have a specific eating behavior such as *organic*, local, ethical food, a diet or *consumption* reduced to ready-made meals.

These eating behaviors are influenced by many factors, both individual (physiological factors, learning, cognitions and environmental. Among the latter, we can mention, as an illustration, the food offer, the media… or social factors, such as, for example, the people who surround us during meals. Indeed, a food behavior can be acquired and developed through social learning. The role of peers and parents has been highlighted.

1. https://stats.oecd.org/Index.aspx?datasetcode=TIME_USE.

2. https://www.lsa-conso.fr/56-des-francais-ont-un-comportement-alimentaire-specifique,331221.

The purpose of this chapter is to examine whether imposed or chosen dietary restrictions can influence "eating together" and to highlight the possible consequences. Can certain dietary restrictions be respected when meals are eaten together? If so, are we not restricting the freedom of those around us?

Several situations must be distinguished. On the one hand, some people are obliged to follow a diet for health reasons, which can make it very difficult to eat certain meals together. In this case, the diet is endured. On the other hand, other individuals have made the choice to pay attention to their diet because of the possible negative consequences of bad food choices (in the broad sense, whether it is on their health, on the planet, on animals, etc.). In this case, the diet is chosen, and it is thus assumed that they can be more flexible with respect to their diet. Is this really always the case? These two categories of individuals will be studied in turn.

In order to answer the questions posed, the dietary constraints endured or chosen will first be presented through two key concepts: nutritional compliance and consideration of future consequences (section 3.2). Then, the case of meals shared between sick and healthy people will be addressed, through a qualitative study (section 3.3). Finally, the situation of meals shared with people who have dietary constraints out of conviction will be analyzed through a second qualitative study (section 3.4).

3.2. Dietary constraints, whether endured or chosen

In this section, we will look at the constraints endured or chosen through two cases. We will first present what the dietary constraints may be, and then we will study the situation of the chronically ill to determine if they can easily eat with healthy people. Finally, we will present the consideration of future consequences, which will influence some food choices in the case of chosen constraints.

It is essential to focus on the situation of the chronically ill, as the number is tending to increase, particularly in France. Indeed, chronic diseases affect 15 million people in France (about 20% of the population) and account for 70% of deaths. However, by eliminating the three risk factors (unbalanced diet, sedentary lifestyle, smoking), 80% of cases of vascular heart disease, stroke and type 2 diabetes could be avoided; the same is true for more than 40% of cancers[3]. In this respect, diet has emerged as one of the solutions: in 2003, the WHO reported that "diet, nutrition and prevention of chronic diseases" demonstrated the direct links between diet and chronic diseases. However, more than 80% of patients suffering from a chronic

3. World Health Organization: http://www.who.int/topics/chronic_diseases/fr.

disease do not follow the therapeutic indications, particularly nutritional indications, to obtain optimal benefits (Lamouroux *et al.* 2005).

Adherence, particularly nutritional adherence, is thus a public health issue insofar as it reduces morbidity and mortality factors[4]. It can be defined as "the degree to which the patient's behavior (in relation to taking treatment, following diets, or making other lifestyle changes) is consistent with medical or health advice" (Haynes *et al.* 1979, p. 1).

The question then arises as to whether the people around a patient can help with nutritional compliance, or whether, in some cases, those close to the patient can unintentionally lead them to non-compliance. Indeed, a diet constrained by a chronic disease can lead to difficulties in eating with people who do not have these constraints. After addressing the issue of dietary constraints, the variables that explain these difficulties in "eating together", as well as the effects on eating behaviors when meals are eaten together, will be presented.

We will then extend our discussion to "chosen" food, through the notion of consideration of future consequences (CFC). Consideration of future consequences is defined as "the extent to which individuals consider the potential distant outcomes of their current behaviors and the extent to which they are influenced by these potential outcomes" (Strathman *et al.* 1994, p. 743). In the case of eating, these are individuals who impose "dietary constraints" on themselves because they are aware of the consequences their eating may have on them or others. This results in restrained eating behaviors on their part according to the consequences they take into account (e.g. veganism, organic food, weight loss diet). Based on these observations, why and how does a person go against the dietary principles in a situation where they have to eat with others? Similarly, to what extent will an individual who does not impose dietary restrictions on themselves comply with the dietary constraints of others when eating together? What about the other's perceived tolerance with regard to oneself?

3.2.1. *What are the possible dietary constraints?*

Food is a necessity, as well as a pleasure. It turns out that the health dimension increasingly occupies a central place in food choices, in particular because consumers are becoming more informed. More recently, other factors have gradually been taken into consideration.

4. Ministère de la santé (France): www.sante.gouv.fr.

In fact, according to the Cetelem observatory, in 2020, the French will be more concerned about improving food production and will be aware of their role as actors. A proportion of 87% of the people questioned in a survey believe that eating better means paying attention to the effects of the products consumed on society and nature. The connection between food and sustainable development seems to have consequences on the food choices of the French. Thus, 80% of consumers prefer local products, some have become vegetarians or flexitarians[5], consume only organic products or have set other dietary constraints, such as a gluten-free or dairy-free diet.

Some people may even be in the process of restricting their food intake, defined as "the tendency to control food intake consciously in order to prevent weight gain or to promote weight loss" (Tuschl 1990, p. 105). This definition emphasizes the notion of "control", which constitutes a cognitive effort to fight the natural urge to eat in order to achieve the goal set by the individual. Thus, in the absence of a pathology requiring a specific diet (e.g. a food allergy), some individuals follow a diet with the objective of losing, controlling or maintaining a weight that represents the standard to be reached (Rideout and Barr 2009). One way is to reduce energy intake.

3.2.2. *Nutritional compliance and the role of social support in chronic disease*

Researchers who have been interested in the issue of chronic disease talk about adherence. Adherence was defined in section 3.1. Some authors refer to adherence as "the degree to which the patient's behavior corresponds to the recommendations agreed upon with the health professional" (Horne 2006). This would be a way of involving the patient more actively and making the patient accountable. In this chapter, we will use the terms "compliance" and "adherence" interchangeably.

Compliance (or adherence) has several characteristics:

– it is conscious or unconscious;

– it is intentional or unintentional;

– it varies over time and may conflict with everyday practice.

5. A flexitarian limits their meat consumption, without being a vegetarian.

In contrast to compliance, non-compliance is a lack of adequacy between patients' behaviors and medical prescriptions regarding diet, exercise or medication (Morris and Schultz 1992).

Adherence is most often studied in the context of medication use. However, it is also linked to other behaviors such as smoking, dietary changes and exercise. In the vast majority of cases, researchers have tried to understand the levers. Their results differ. This can be explained by differences in the choice of adherence measurement (direct or indirect methods)[6], as well as by the variability of the chronic diseases studied.

Some explanatory factors for non-compliance have been identified (Reach *et al.* 2015):

– individual characteristics: socio-demographic and cultural characteristics, psycho-sociological and anthropological factors, motivations, knowledge of the disease, beliefs and attitudes towards the disease, perception of risk, severity of the disease being treated, mental representations of the drug and the disease and so on;

– the characteristics of the therapy, including the drug protocol, hospital stays, the price of the drug and its reimbursement;

– factors related to health care providers (doctors, nurses, pharmacists, etc.);

– environmental characteristics, including the role of the family and caregivers.

This last element thus shows the role of social aspects, which are particularly important in explaining compliance.

Indeed, the literature has highlighted the importance of social support in the case of illness. It includes the number of friends available to an individual, as well as the satisfaction that this social support brings (Sarason *et al.* 1983). It therefore refers to the comfort, care, esteem or help given to an individual by those around them.

6. For example, to determine whether a patient with coronary artery disease is adhering to their treatment, drug metabolites and concentrations in the patient's blood or urine can be studied. Indirect methods include questionnaires, self-reporting, remaining pill counts, prescription refill frequency and electronic pillboxes. Punctuality to appointments can be a measure, although it is not widely used.

Several typologies of social support exist. In particular, two types of social support can be distinguished (Lett *et al.* 2005):

– structural support (or support network), that is, the type, size, density and frequency of contact with the network of people available to an individual;

– functional support, which corresponds to the perception of the benefit provided by this structure.

Many studies have examined the influence of social support on health. Social support is most often seen as beneficial as the absence of social support is itself a source of stress. However, some research has suggested that this effect may be negative (Hughes and Curtis 2000). For example, the role of social support in predicting a patient's post-myocardial infarction quality of life has been shown; perceived rather than actual support has a greater influence. Because its effects are not always positive, it is important to understand its role in the case of shared meals and nutritional compliance.

As we have seen, there is much less research on nutritional adherence than on medication adherence. Yet, a patient adheres more readily to their medication than to their diet. According to Mbaye *et al.* (2013), the factors that explain medication adherence (level of knowledge of the disease, severity of the disease, level of health coverage) are not quite the same as the factors influencing diet adherence (only knowledge of the disease is significant in this case).

Because of the lack of research on nutritional compliance and the frequency of non-compliance in this area, it was important to investigate the role of the close circle (i.e. social support) on this variable: is a restricted diet a barrier to "eating together?" What role can those closest play in nutritional compliance? Is this effect positive or negative? These questions will be answered in the second section of this chapter.

3.2.3. *Selected dietary constraints: the example of consideration of future consequences (CFC)*

We have previously shown that an individual can endure dietary restrictions for health reasons, mainly in the case of a chronic disease. However, some individuals without health problems can impose similar dietary constraints on themselves (eliminating or limiting certain foods in favor of others deemed better for their health or the environment, and consequently, the health of others in the long term). Thus, an individual may adopt a certain diet, considering the future consequences that may result from their action. According to the literature (Strathman *et al.* 1994; Orbell and Hagger 2006; Adams and Nettle 2009), each individual would tend to be concerned about the future consequences of their actions to varying degrees. This phenomenon is thus called consideration of future consequences (CFC – for a definition, see section 3.2). As in the case of compliance, discussed above, social aspects should not be neglected when studying CFC. Indeed, despite its relatively stable nature, some authors indicate that CFC is a psychological disposition that can evolve according to the individual's life environment, particularly their social environment (Demarque *et al.* 2010).

The behavioral consequences of CFC have already been studied in the literature. Considering the future consequences of our actions has been shown to reduce undesirable behaviors, such as smoking (Adams and Nettle 2009), and to increase a number of desirable behaviors, such as a preference for public transportation (Strathman and Joireman 2005; Joireman *et al.* 2006).

With respect to diet, however, the benefit of strong CFC is less clear. According to Adams and Nettle (2009), the greater the consideration of future consequences, the lower the body mass index (BMI)[7]of individuals, but the authors do not establish a link with diet. Joireman *et al.* (2012) show that preference for healthy eating would be influenced by CFC. However, this research does not address the social aspects of eating, the effect of CFC on communal eating. Yet, the social aspects related to the conviviality of meals are more central to French culture than to the English-speaking culture in which the main studies on CFC have been conducted. It is therefore legitimate to ask whether individuals who do not attach importance to the consequences of their food eat easily with other individuals for whom the future consequences of their food are essential and have transformed their eating habits. This variable can be measured using a scale called the "Consideration of Future Consequences" by Strathman *et al.* (1994).

However, this scale can be improved and does not take into account the situational aspects of eating, particularly meals eaten together. Only qualitative studies can help us understand how a chosen diet (in consideration of future consequences) is a barrier to "eating together?" What role can the environment play in whether or not an individual respects the constraints that they voluntarily impose? Is this effect positive or negative, both for the person imposing these dietary constraints and for the individuals with whom they share a meal? These questions will be answered in the third section of this chapter.

---

7. BMI is the ratio of weight in kg to height in cm.

1. I consider how things might be in the future and try to influence those things with my day-to-day behavior.

2. Often I engage in a particular behavior in order to achieve outcomes that may not result for many years.

3. I only act to satisfy immediate concerns, figuring the future will take care of itself.

4. My behavior is only influenced by the immediate (i.e. a matter of days or weeks) outcomes of my actions.

5. Convenience is a big factor in the decisions I make or the actions I take.

6. I am willing to sacrifice my immediate happiness or well-being in order to achieve future outcomes.

7. I think it is important to take warnings about negative outcomes seriously even if the negative outcome will not occur for many years.

8. I think it is more important to perform a behavior with important distant consequences than a behavior with less important immediate consequences.

9. I generally ignore warnings about possible future problems because I think the problems will be resolved before they reach crisis level.

10. I think that sacrificing now is usually unnecessary since future outcomes can be dealt with at a later time.

11. I only act to satisfy immediate concerns, figuring that I will take care of future problems that may occur at a later date.

12. Since my day-to-day work has specific outcomes, it is more important to me than behavior that has distant outcomes.

The CFC scale is formulated so that a higher score indicates greater consideration of future consequences. To do this, items 3, 4, 5, 9, 10, 11 and 12 must be recoded in reverse order at the time of statistical processing.

Box 3.1. *CFC measurement scale, Strathman* et al. *(1994)*

3.3. Suffering from dietary constraints but eating with others: the case of meals between sick and healthy people

Can people with chronic illnesses easily eat with others? In other words, do shared meals have an influence – either positive or negative – on nutritional compliance? It seemed important to us to have the point of view of the persons concerned in order to determine their particular constraints and what difficulties they may encounter.

The objective of this study was to find out whether chronically ill patients are dietary compliant (whether they are able to follow the diet recommended to them) and to identify the factors that make it easier for them to be compliant. In particular, questions were introduced on the role of those closest, in order to determine whether the chronically ill can easily eat with other people.

Semi-structured interviews were conducted until the threshold of semantic saturation was reached.

Thus, 17 people (7 men; 10 women) aged between 21 and 77 years and suffering from a chronic disease (diabetes, hypertension, fibromyalgia, Crohn's disease, food allergies, etc.) were interviewed.

Box 3.2. *Methodology of Study 1*

First name	Gender	Age	Pathology	Profession	Family situation
Annie	F	70	Cholesterol and hypertension	Retired	In a relationship and has a child
Audrey	F	35	Allergy	Commercial assistant	In a relationship and has a child
Benjamin	M	23	Allergy	Student	Single, no children
Claudette	F	77	Diabetes	Retired	In a relationship and has a child
Flavienne	F	61	Allergy	Home Help	In a relationship and has a child
Gaétan	M	22	Diabetes	Unemployed	Single, no children
Gabriel	M	45	Hypertension	Theater stage manager	In a relationship but no children
Grégoire	M	21	Obesity	Student	Single, no children
Jacques	M	76	Cholesterol and hypertension	Retired	In a relationship and has a child
Léa	F	25	Crohn's disease	Activity leader for children	Single, no children

First name	Gender	Age	Pathology	Profession	Family situation
Marielle	F	52	Obesity	Teacher	Single with a child
Nicole	F	46	Fibromyalgia	Nursery assistant	In a relationship and has a child
Noémie	F	32	Crohn's disease	Unemployed	Single, no children
Philippe	M	53	Cholesterol	Engineer	In a relationship and has a child
Suzanne	F	64	Diabetes	Auxiliary nurse	In a relationship and has a child
Valérie	F	49	Diabetes	Nurse	In a relationship and has a child
Yann	M	53	Hypertension	Computer scientist	In a relationship and has a child

Table 3.1. *Participants in Study 1*

According to the results of this study, a diet is generally difficult to follow. Deviations from the diet may occur for several reasons. When sick and healthy people plan to eat together, the difficulties in following the diet are usually accentuated. However, this depends on the context, the people involved and the disease concerned: some diets are difficult to follow, regardless of the people with whom the meals are taken. For other respondents, the diet is more or less difficult to follow depending on the people around them.

3.3.1. *The difficulties of the system*

In general, dieting is often seen as restrictive and difficult for respondents, whether they are eating alone or in groups: "I always eat the same foods (potatoes, rice, boiled pasta with nothing on it, white meat, fish and rusks instead of bread), until my body gets better and allows me to deviate. In any case, it is very difficult. Now I hardly feel any pleasure in eating, except when I overindulge" (Noémie). "Finally, I would have to be on a desert island with only fruits and vegetables around me! On a more serious note, my diet is quite difficult to follow since many foods are forbidden to me" (Nicole).

For example, some of them say that they often deviate: "In the summer, seeing my friends and family at the table while enjoying a barbecue, French fries, drinking rosé during the aperitif... It was really hard and I unfortunately sometimes cracked; it was stronger than me" (Nicole).

On the other hand, a few people find that it is not complicated to respect their diet. This is the case for Benjamin, for example, because he has been used to it since he was a child: "Since I was one year old I have always eaten like this, so I am used to it." If he does not respect these dietary constraints, he risks angioedema (he has already had three). According to him: "Everyone knows and is used to it, so I am supported by my nearest and dearest."

In addition to the difficulty of following the diet, there are other reasons for deviations.

3.3.2. *Factors that explain deviations from the plan*

Deviations from the diet depend first of all on the type of illness, the pathology. Some people tolerate deviations, others do not: "You have to be mentally strong to follow the diet in an irreproachable way. As one of those people who do deviate, I understand that there are deviations. However, when our health requires it, it is essential to do whatever it takes to get back on track" (Noémie).

People who have deviated from their diet say that they can easily return to it afterwards. However, it should be noted that this depends on the type of condition: in some cases, the diet must be followed very strictly. For example, according to Suzanne, "If I don't follow my diet properly, my blood sugar can rise and I can go into a diabetic coma. I am risking my life".

The length of time that the disease, and therefore the diet, has been present can also have an influence on the frequency of deviations. For example, Benjamin, who has food allergies, says he follows his diet "every day, every week, all year long". It is therefore important that the support of those around him be present throughout the illness.

This difficulty in following the diet may be accentuated at certain times. Thus, the diet may be more difficult at certain times and easier at others. This is the case, for example, during holidays: "During holidays or during family meals when there are things on the table, when there are many of us, it certainly makes you want to" (Annie). In this case, in addition to the particular moment concerned, the family involuntarily plays a role in the failure to respect the diet.

For several respondents, summer is a source of additional difficulties: "The hardest time to follow my diet is in summer. There are too many temptations, both in terms of food and alcohol!" (Grégoire). This is also the case for Nicole, or Suzanne: "Summer is synonymous with good times, good aperitifs and good meals…" (Nicole). "For me this diet is harder to follow in summer because I love fruit and I have to limit myself" (Suzanne). Summer is therefore synonymous with pleasure shared with family or friends, as well as with deviations from the diet.

3.3.3. *Meals with other people: a variety of situations*

In general, respondents say that when a meal is shared with others, it has a negative impact on the diet followed:

> "Either I avoid going out at meal times or I bend my diet" (Noémie). Benjamin said that he is forced to eat differently in this case, "unless the dishes don't have the discouraged foods."

We change our eating behavior when we are in a group:

> In a friendly atmosphere, we take more time to eat, we tend to eat more. To honor the dishes and also because we add what is not on the diet, like wine for example, or we eat things like chips, which I only eat when I am with friends (Philippe).

Thus, close relations can hinder the diet:

> Family or friends don't help me at all. The challenge is not to bother others with your food choices. But for me, it's not worth seeking help from those around me; sometimes they taunt me (Gabriel).

On the contrary, in other cases, the people around the person can help them to follow the diet:

> My close friends and family insist a lot that I follow what is written in the guide and helps me by proposing varied menus, by reducing

> access to certain products in the fridge while maintaining certain things that give pleasure (Philippe).

Here we can see the beneficial role of social support, which indirectly contributes to the good health of the chronically ill.

However, it depends on the people involved: some people have a more beneficial role than others.

Among the people around them, family is, in most cases, a great help, especially at the beginning of the disease. They allow the patient to adapt to new constraints and help them modify their eating habits. This support from the family is sometimes also manifested later, but in a more moderate way, during the different meals, through advice. It is even more important if another person in the family has to follow a diet for health reasons: it increases motivation.

> My parents were very supportive when I found out about my illness, which was a great relief. My mother was the most supportive and helpful. She made a lot of effort to make sure I was comfortable at the table. This helped me to follow my diet without missing family meals and isolating myself (Léa).

> Those closest to me were very present at the beginning of my diet, especially my family. I'm thinking especially of my son, who was particularly keen to help me and therefore followed the same diet as me for the first few days. From now on, during my weak moments, my family and friends encourage me not to crack. They are always close to me and do everything they can not to tempt me (Nicole).

We can see that among the people around us, the family is cited first.

On the other hand, friends or professional relations do not play the same role: according to the respondents, dieting is more difficult to follow during meals with friends or in the professional context. This is because the food often does not correspond to what is prescribed in the case of a diet.

> When we go to eat at a friend's house, it will be pizza, aperitif (Valerie).

In fact, several times during the interviews, meals with friends were considered the most complicated.

> The most difficult time to follow my diet is when I'm at a party with friends who don't have the same habits as me. Not at restaurants because everyone orders what they want, but when there's a dinner that's prepared for people who have a general, well, normal diet. So I always get little comments like, do you eat that? I feel like I'm in the way during those times (Gabriel).

A similar situation is found in the case of professional lunches. According to Léa, "When I find myself at a lunch with acquaintances, professional relations or simply friends, I can see that my eating habits differ from those of others; it is sometimes obvious. What would help me feel better about this diet is if the people around me would change their eating habits a little and eat more healthily". For Léa, the people around her could play the role of social support and help her to stick to her diet.

We can see that sick and healthy people can eat together. For a person with a chronic illness, the diet is easier to follow alone, or when meals are with close family members, especially during daily meals. The difficulty increases when the individual is in the presence of other people around them: friends or professional relations, for example. The latter do not play the role of social support. On the contrary, their influence can lead to non-compliance with the diet.

What about those who do not have a chronic disease but impose dietary constraints on themselves out of awareness of the future consequences of their diet?

3.4. Having dietary constraints out of conviction: How do you eat with others?

A second study allowed us to better understand the complexity of communal meals among people with dietary constraints based on conviction (consideration of future consequences). The methodology detailed in Box 3.3 allowed us to compare the points of view of those who claim to have dietary constraints through choice and those who declare having a diet that can be described as "classic", without any particularity. In both cases, difficulties are felt in "eating together". Those who have dietary constraints feel obliged to justify themselves, but cope with the situation by trying to convince others that their food choice is the right one. Those who do not have constraints are more apprehensive about sharing meals, which they see as a headache, wanting to please without feeling judged by those with strong convictions in favor of alternative food.

Participants and study process

Three filmed and transcribed focus groups (duration: 2 hours 15 minutes to 2 hours 50 minutes each) with varied socio-demographic profiles (28 individuals aged 22 to 65 years) and eating behaviors.

Focus group No. 1: activists (10 participants) – Mainly made up of "activists" for alternative food (vegan, gluten-free, organic, etc.)

Focus group No. 2: classic (9 participants) – Consisting of individuals declaring to have a "classic diet, without particularity"

Focus group No. 3: mixed group (9 participants) – Made up of a mixture of the two types of profiles mentioned above

The discussion guide was constructed as a funnel, starting with general questions about eating habits (shopping, favorite products, meals, etc.), followed by the importance placed on diet, and then to the consequences and finally to whether or not these consequences were taken into account.

Analysis

The transcribed focus groups were analyzed using NVivo 11 software. Coding was done by two researchers and verified by a third. The tree structure was built from the main themes of the discussion guide for the parent nodes, then in vivo from the field data for the child nodes (as close as possible to the verbatim). Initially, the objective of these focus groups was to explore the notion of consideration of future consequences mentioned in section 3.2.2. The theme of "eating together" emerged from the field and was thus coded as a theme in its own right. It is only this theme that will be discussed below (sections 3.4.1 and 3.4.2).

Box 3.3. *Methodology of Study 2*

3.4.1. *The point of view of those who impose constraints on themselves: wanting to convince without being judged*

Those who are constrained by conviction, because of a strong consideration of the future consequences of their food (on health, the environment or animal welfare), essentially share a desire to convince their loved ones, in particular, to pass on their vision of food to their children.

> Passing on something to our children, for example, if we get them used to eating healthy, organic, local food, they won't go and eat at McDonald's, Quick and all that. I see my children eating everything and things that other children do not eat (Nicolas 2).

> So, since she was a little girl, I got her used to eating well, she never drinks soda, she doesn't eat candy or chocolate except for organic ones (Romain).

This alternative diet sometimes originates from the convictions of the parents of the interviewees and thus continues from generation to generation:

> I think that human beings should not consume cow's milk, it is simply not good for us. It's not digestible, it's an aberration, it's my mom who introduced me to it. She didn't consume a lot of it and at the beginning of the 2000s she showed me a little book with the effect of milk and stuff, I leafed through it as a teenager as I didn't care, then as I grew up I became aware of it (Djibril).

However, the intergenerational transmission is sometimes reversed, people with dietary constraints out of conviction try to convince their parents to change their behavior:

> I did it the other way around, I gave my mother a taste for vegetable gardening, you have to give them a hand and then they do everything on their own (Romain).

If it is important for these people to convince their close circle, some go further and do not miss the opportunity to advocate for an alternative diet with people they do not necessarily know. Romain, who participated in the "mixed" focus group, tried several times to convince Marie-Laure, a participant who defined her diet as "classic, without any particularity".

> (Marie-Laure): "Aspartame that replaces sugar is another sugar."

> (Romain): "It is better to take a product with Stevia, it is better."

> (Marie-Laure): "Yes, I know, but I still consume, after the fats I don't consume too much."

> In the "activists" group, participants also took advantage of the focus group to share their tips with each other:

(Arnaud): "How do you make your own apple juice?"

(Djibril): "With a juice extractor."

This diet, dictated by a very strong awareness of future consequences, leading to particular diets (vegetarianism, veganism or even being bio-exclusive), generally comes up against the gaze of those around them, or at least of the people they are unable to convince. In spite of strong convictions, some respondents suffer from misunderstandings with those around them, not understanding why they impose so many constraints on themselves.

> It's very hard to get people to listen to me; I share my life with someone who doesn't share my choices at all, who is a meat eater. This is one of the things that means we are on the verge of breaking up because we don't go in the same direction. We have to run specific errands every time, justify ourselves because the other person asks why. In-laws who don't fit in. Many people don't understand these choices. We are very quickly judged (Maëva).

Confronted with comments that they consider extreme or based on prejudices ("Once, in relation to soya milk, I was told ah you eat cardboard [...] In fact, they have preconceived ideas, negative ideas about it" – Véronique), they would like to convince people so that their way of eating becomes the norm, but they are faced with people who are not very receptive, whom they see as preaching an industrial diet:

> Food is the same, if you eat organic food, people say to you but why do you do that, you pay a lot. People judge you quickly because you are not like them and they want you to be (Nicolas 2).

The "norms" of each person then come into conflict. The dietary constraints are perceived as so strong that the vocabulary used by relatives is that of illness, of disadvantage, as if by choosing an alternative diet, the individual was disadvantaging themselves:

> I remember a friend, when I told her what I eat and don't eat, who said to me: 'but it must be a disadvantage in everyday life' [You don't experience it like that at all?] Not at all, it's like it's not natural not to eat Nutella when it's the other way around! (Nicolas 1).

They would like the diversity of diets to become part of the customs, to share their tastes without being considered outside of normality, sick or disadvantaged, like José, who makes the parallel with music:

> It's strange because in other areas it doesn't apply. I mean, for example, I have a friend who doesn't listen to music, we're not going to call her disadvantaged, but when it comes to food, it crystallizes people (José).

3.4.2. *The point of view of those who do not have constraints: wanting to make an effort but not too much*

The impression of being judged by others appears more in the background among respondents with a more traditional diet and appears to be linked to particular situations, when they are invited to the homes of people with dietary constraints or when they have to invite them:

> The vegans will criticize everything in my home, even my sister who watches a lot of what she eats and who also makes remarks (Benoît).

However, alternative food activists believe they are tolerant and do not criticize their hosts:

> If I eat at a friend's house and the vegetables are not organic, I won't make a fuss or call them polluters (Arnaud).

Thus, among people who have no constraints, it is not so much the criticism that bothers them but the impression that it is a hassle to get everyone to eat the same thing, to not be judged as well as to have a certain conviviality, a pleasure of eating together.

> When I was a student, we used to have big parties, there was sausage, there was everything, Jewish, Muslim, Christian, and nobody would have thought about the fact that there was sausage, whereas nowadays we have to be more careful, I surprised myself by buying halal sausage because otherwise he would spend the evening eating cheese, I think you have to take care of your guest, I made this decision but I was very surprised at myself for making the decision (Vincent).

The people who do not have any dietary constraints then express their difficulties in pleasing everyone. "I have a friend of my mother's who is a headache in the sense that she hardly eats anything, so when my mother wants to cook for her, it's complicated, she has to ask what to make for her, so it's a bit annoying, so we have to adapt, there's a social side" (Aurélie). The fact that the dietary constraints of their guests are chosen and not endured because of a health problem is sometimes difficult to understand. According to them, it would be simpler if their guests would accept making an occasional deviation from their principles:

> I'm a bit of a pain, but if you can't eat that when there's no problem, it's annoying (Valérie).

However, sometimes, even if the guest with dietary constraints out of conviction does not openly criticize their hosts, they refuse to eat certain products:

> I didn't say it before, but I don't buy any processed products, so if I want to make a fruit juice, I buy the fruit and I make it myself. I never deviate from this, anything that is prepared or processed I don't buy and if there is any at my friends' house I don't consume it either (Djibril)

which can be perceived as offensive. Some also deplore one-sided efforts, in the sense that when it is the turn of the person with an alternative diet to entertain, they generally want to introduce others to their diet and will not prepare a more conventional dish, imposing their vision of food:

> I have a colleague who told me that she had been invited to the home of raw food eaters who eat sprouts and all that, and when she arrived there was only that. It's complicated this notion of conviviality, you have to put yourself in the place of the other person and it's rare now (Véronique).

3.4.3. *Struggles, trade-offs and compromises of eating together*

Finally, this comparison between people with dietary constraints out of conviction (consideration of future consequences) and those without reveals an awareness of the difficulties of eating together, especially at home, or to a lesser extent, at work, due to the pressure of the majority group:

> The motto of the service is: fat is life, so that means everything, so it's coffee breaks, aperitifs, hamburgers, pizza. It's very problematic because there's the group effect, so either you have the strength to fight or you don't and I don't (Véronique).

Some people try to find more or less extreme solutions according to the constraints of their diet so as not to deviate from their principles in these situations. For those who have a classic diet, without any strong particularity, it is simply a matter of arbitrating to choose the product judged to be the "least unhealthy":

> Afterwards, if I go to a friend's house, I'll choose what is least bad for me, between a diet Coke or a non-diet Coke (Benoît).

For those with strong dietary constraints, the choice is between a single dish for everyone or different dishes. Choosing a single dish can sometimes lead to lying about its composition, to avoid criticism.

> I saw this at Christmas, we invited friends to the house and all that. I made vegetarian lasagne, without telling anyone that there was no meat in it. Everybody was surprised by the dish, they said: "but do you eat meat?" I told them "yes, it's Christmas." But if I had told them right away that there was no meat, there would have been comments like "ah well yes, you can smell that there is no meat, yes it's not good" (Héléna).

Choosing to bring your own individual portion can sometimes ease tensions and focus on being together, regardless of the food.

> I know that the people around me understand. When I see my parents and I tell them I'm starting my gluten-free phase, they do the shopping accordingly. Afterwards, depending on where I am going, I take my Tupperware, the idea is to have a good time together, not to want what's on the neighbor's plate (Véronique).

Finally, the restaurant (at least when there is a consensus on the choice of the establishment) appears to be the place where we can eat together without stress, because the diversity of dishes and restaurants means that everyone can, with some exceptions, find something suitable for their diet. The confidence in the quality and origin of the products also makes some people relax, who are less careful with their principles:

> No, when we go to a restaurant it's to be served, to relax, to put our feet under the table (Didier).

> When you go to a restaurant and they serve good meat, you already know it's going to be quality (Christine).

3.5. Conclusion

The objective of this chapter was to see how dietary restrictions, whether imposed (due to chronic illness) or chosen (based on whether or not we consider the future consequences of our diet), can influence "eating together". To address this objective, two qualitative studies were conducted.

The results of the first study show that when sick and healthy people plan meals together, difficulties in following the diet are generally accentuated, especially on festive occasions such as Christmas, or during certain seasons, such as summer.

However, it depends on who is with us at the meal. While there are many reasons for deviating from the diet, sharing a meal, especially with friends or business associates, usually increases the difficulty of following the diet. In this case, the sick and the healthy can eat together, but this can have negative effects on the eating behavior of the sick.

This first study also shows that, among the people around them, the family often has a beneficial effect and plays the positive role of social support. This is particularly true at the beginning of the disease. This social support role of the family is reinforced when another person in the family is also on a diet, whether this choice is voluntary or imposed. In this case, when sick and healthy people "eat together", the consequences are positive for everyone.

The second empirical study is devoted to "desired" dietary constraints, that is, less restrictive. In this case, we see a desire on the part of people who have made these dietary choices to convince the people around them to adopt the same type of behavior. They sometimes even take on the position of activist, always trying to convince, even those they do not know well. This was not evident in Study 1, as the diet was associated with a pathology and therefore not justified outside of this case.

However, in both cases, the dietary constraints may come up against the gaze of those around you, who are sometimes surprised by the diet adopted. This can happen in particular in the case of specific diets (vegetarianism, veganism or bio-exclusive). The dietary constraints are perceived as so strong that the vocabulary used by relatives is that of disease. The results of the two studies are similar on this point.

This comparison between people who have dietary constraints out of conviction (consideration of future consequences) or obligation (chronically ill) and those who do not reveals an awareness of the difficulties of eating together, especially at home or when invited. This is somewhat less true when eating out, since the choices are wider. But even in this last case, the difficulty is increased when meals are taken with friends or professional relations.

The comparison between the two types of dietary constraints, desired and chosen, reveals many common points. "Eating together" is always possible and desirable, even if a reciprocal adaptation of the person who has these constraints and those around them is often necessary.

3.6. Appendix: Characteristics of the Study 2 sample

First name	Gender	Age	Food	Occupation	Family situation
Focus group – "activists"					
Arnaud	M	29	Organic + local	Graphic designer	Single with child
Héléna	F	30	Vegetarian	Agricultural worker	Single, no children
Djibril	M	30	Organic	Company creator	In a relationship, no children
José	M	56	Organic + diet	Trainer for adults	Single, no children
Maëva	F	28	Vegan	Educational assistant	In a relationship, no children
Nicolas 1	M	35	Diet + organic	Insurance advisor	In a relationship, has a child
Nicolas 2	M	24	Gluten free	Clinical psychologist	Single, no children
Nicole	F	52	Gluten free + vegetarian	Bookseller	Single with child
Valérie 1	F	56	Diet	Administrative technician	In a relationship, has a child
Véronique	F	40	Gluten free	Caregiver	Single, no children
Focus group – "classic"					
Aurélia	F	38	Classic	Assistant	In a relationship, has a child
Bénédicte	F	46	Classic	Assistant	Single with child
Benoît	M	36	Classic	Teacher–researcher in computer science	In a relationship, no children
Christine	F	46	Classic	Director PMGE	Single with child

Didier	M	57	Classic	Security agent	Single, no children
Jacques	M	65	Classic	Retired	In a relationship, has a child
Marie-Claude	F	52	Classic	Head of student life services	In a relationship, has a child
Valérie 2	F	45	Classic	Competency development manager	In a relationship, has a child
Vincent	M	53	Classic	Elementary school principal	In a relationship, has a child
Focus group – "mixed"					
Amelle	F	22	Classic diet trend	Education student	In a relationship, no children
Anne-Sophie	F	30	Classic diet trend	Optician	In a relationship, no children
Anthony	M	26	Vegetarian	Composer	Single, no children
Antoine	M	28	Classic	Teacher–researcher in information systems	Single, no children
Isa	F	49	Vegan	Life coach	In a relationship, no children
Manuel	M	52	Organic + local	Casual entertainment worker	Single, no children
Marie-Laure	F	31	Classic	Administrative manager	Single with child
Nicolas 3	M	38	Gluten free + organic	Technical officer	In a relationship, has a child
Romain	M	39	Organic	Farmer	In a relationship, has a child

3.7. References

Adams, J. and Nettle, D. (2009). Time perspective, personality and smoking, body mass, and physical activity: An empirical study. *British Journal of Health Psychology*, 14, 83–105.

Bellisle, F. and Dalix, A.M. (2006). Modifier son comportement alimentaire : mission impossible ? *Cahiers de Nutrition et de Diététique*, 41(3), 159–165.

Demarque, C., Apostolidis, T., Chagnard, A., Dany, L. (2010). Adaptation et validation française de l'échelle "Considération of Future Consequences". *Bulletin de Psychologie*, 63(5), 351–360.

Haynes, R.B., Taylor, D.W., Sackett, D.L. (1979). *Compliance in Health Care*. John Hopkins University Press, Baltimore.

Horne, R. (2006). Compliance, adherence, and concordance: Implications for asthma treatment. *Chest*, 130(1), 65S–72S.

Hugues B.M. and Curtis R. (2000). Quality and quantity of social support as differential predictors of cardiovascular reactivity. *Irish Journal of Psychology*, 21, 16–31.

Joireman, J., Strathman, A., Balliet, D. (2006). Considering future consequences: An integrative model. In *Judgments Over Time: The Interplay of Thoughts, Feelings, and Behaviors*, Sanna, L.J. and Chang, E.C. (eds). Oxford University Press, Oxford.

Joireman, J., Shaffer, M.J., Balliet, D., Strathman, A. (2012). Promotion orientation explains why future-oriented people exercise and eat healthy: Evidence from the two-factor consideration of future consequences – 14 scale. *Personality and Social Psychology Bulletin*, 38(10), 1272–1287.

Lamouroux, A., Magnan, A., Vervloet, D. (2005). Compliance, therapeutic observance and therapeutic adherence: What do we speak about? *Revue des maladies respiratoires*, 22(1), 31.

Lett, H.S., Blumenthal, J.A., Babyak, M.A., Catellier, D.J., Carney, R.M., Berkman, L.F., Burg, M.M., Mitchell, P., Jaffe, A.S., Schneiderman, N. (2007). Social support and prognosis in patients at increased psychosocial risk recovering from myocardial infarction. *Health Psychology*, 26(4), 418.

Mbaye, A., Ntontolo, F.L.K., Diomou, A.F., Bodian, M., Ndiaye, M.B., Kane, A., Yaméogo, N.V., Pessinaba, S., Sarr, S.A., Dioum, M. (2013). Prévalence et facteurs liés à l'observance thérapeutique chez des patients noirs africains coronariens stables, suivis en ambulatoire en service de cardiologie de Dakar au Sénégal. *Annales de Cardiologie et d'Angéiologie*, 62(1), 17–21.

Morris, L.S. and Schulz, R.M. (1992). Patient compliance – An overview. *Journal of Clinical Pharmacy and Therapeutics*, 17(5), 283–295.

Orbell, S. and Hagger, M. (2006). Temporal framing and the decision to take part in type 2 diabetes screening: Effects of individual differences in consideration of future consequences on persuasion. *Health Psychology*, 25(4), 537–548.

Reach, G., Meynier, D.G., Darné, B., Herpin, D. (2015). Facteurs associés à la non-observance chez les hommes et chez les femmes hypertendus non contrôlés. *Annales de cardiologie et d'angéiologie*, 64, 222–226.

Rideout, C.A. and Barr, S.I. (2009). "Restrained eating" vs. "trying to lose weight": How are they associated with body weight and tendency to overeat among postmenopausal women? *Journal of the American Dietetic Association*, 109, 890–893.

Sarason, I.G., Levine, H.M., Basham, R.B., Sarason, B.R. (1983). Assessing social support: The social support questionnaire. *Journal of Personality and Social Psychology*, 44(1), 127.

Strathman, A. and Joireman, J. (2005). *Understanding Behavior in the Context of Time: Theory, Research, and Application*. Lawrence Erlbaum Associates Publishers, Mahwah.

Strathman, A., Gleicher, F., Boninger, D.S., Edwards, C.S. (1994). The consideration of future consequences: Weighing immediate and distant outcomes of behavior. *Journal of Personality and Social Psychology*, 66(4), 742–752.

Tuschl, R.J. (1990). From dietary restraint to binge eating: Some theoretical considerations. *Appetite*, 14, 105–109.

4

Eating Together, Yes, But Without Meat! Social Influences Related to Vegetarianism and Veganism

Gilles SÉRÉ DE LANAUZE[1], Lucie SIRIEIX[2] and Erick SUAREZ DOMINGUEZ[1]

[1] *MRM Laboratory, University of Montpellier, France*

[2] *UMR MoISA, Institut Agro-Montpellier SupAgro, France*

4.1. Introduction

Judging from the multiplication of press articles, exchanges on social networks and the offer of dedicated products on shelves and restaurant menus, vegetarianism is in fashion. In fact, nearly 2% of the French population declare themselves vegetarians, and 0.5% vegans[1], which is still a few tens of thousands of people within the French territory. Like other food consumption trends that are currently experiencing a resurgence of interest, such as buying organic products and locavorism (consumption of local products), this phenomenon is largely the result of broader concerns and convictions emerging within Western societies. In particular, health, social, environmental and even moral issues are influencing more and more

1. Source: Xerfi, 2020.

Evolution of Social Ties Around New Food Practices, coordinated by Gilles SÉRÉ DE LANAUZE.

individuals' choices and behavioral changes in terms of food and the adoption of particular practices and diets.

While these new practices are still relatively marginal, and are variably accepted by the entourage and the societal environment, they are nonetheless topical phenomena and the subject of much debate. Tensions can arise between practitioners and non-practitioners. The case of vegetarianism is an illustration of this. Vegetarianism benefits from the health and ethical scandals surrounding meat products; it is supported by the growing number of recommendations to reduce meat consumption (notably by the Intergovernmental Panel on Climate Change (IPCC)) and finally, it has the support of actors (vegetarian and vegan) in a rapidly growing market (24% increase in sales in France between 2018 and 2019[2]). We thus observe the adoption of practices excluding meat in a more or less drastic way. Depending on individual motivations and choices, vegetarianism covers a wide range of practices, based in particular on the extent of the range of foods or products excluded and the degree of adherence of the adopted diet. Between the "flexitarian", who reduces their consumption of meat products while tolerating deviations, and the strict vegan, who proscribes all products derived from animal exploitation (including honey, silk and leather), a wide variety of levels coexist, which we will describe later.

Deciding not to eat meat, beyond the change of diet that it implies, modifies habits and social relations linked to food. This decision can have strong consequences on the way we live with others, their views, their reactions and the new place given to the vegetarian. These consequences are perceived by the vegetarians in the society. But the vegetarian is not alone. Sometimes, they are able to count on the group, the community and other vegetarians to facilitate their practice, to overcome feelings of isolation and to reinforce their commitment. These relational and social influences and their effects on eating behaviors are the focus of this chapter. It presents the findings from two qualitative studies drawing on life stories (Bertaux 2016). Participants were recruited by word of mouth, with the aim of ensuring qualitative and not statistical representativeness (i.e. seeking profiles of individuals who may have varied backgrounds with respect to their transition to vegetarianism).

The first study was conducted in 2017 with 25 young adult vegetarians (16 females and 9 males, aged 17–41 with a mean age of 25.1 years) with the primary objective of understanding the process leading them to vegetarianism and their relationships with non-vegetarians. Variety in terms of practices was sought, with the sample consisting of 2 flexitarians, 11 vegetarians, 6 strict vegetarians and 6 vegans.

The second study, conducted in 2019, also focused on the process leading to vegetarianism and people's relationships with non-vegetarians, whilst looking from

2. Source: Xerfi, 2020.

a dynamic perspective at the trajectories of vegetarians and their relationships with non-vegetarians and the vegetarian group. The sample consisted of 7 men and 8 women, with an average age of 33.8 years (min = 18, max = 50). Respondents were vegetarian (9), strict vegetarian (5) or vegan (1) for, on average, 8.9 years (min = 6 months, max = 30 years).

4.2. Not eating meat!

Meat is an important part of Western culture and consumption habits. It is considered an essential and even noble part of the daily menus of a majority of French people. Indeed, France remains a heavy meat consumer, with an average consumption of 87.5 kilos of meat per year and per person[3]. With a gastronomic tradition that favors meat in all its forms, the French are among the largest consumers in the world, although they remain far behind North Americans (120.2 kg), Australians (111.5 kg) and Austrians (102 kg)[4]. However, criticism and warnings are multiplying about the consequences of eating meat that are potentially harmful to health and the environment. Excessive meat consumption raises the risk of cardiovascular diseases and some types of cancer. Meat production is also a major cause of pollution and a waste of resources, particularly water and agricultural land, and thus induces numerous misdeeds: pollution, deforestation and animal suffering. Recent consumption data shows a downward trend over the last 10 years, especially for red meat (Lepiller *et al.* 2016).

The critical discourse about meat makes many consumers feel uneasy, due to the contradiction between habits and a taste for meat consumption and an awareness of its harmful effects (Rothberger 2020; Séré de Lanauze and Siadou-Martin 2016). Some people take the plunge and stop eating meat. These "vegetarians" are still in the minority (Edwards 2013), but they are increasingly listened to by the media, society and even public authorities, which are encouraging meat alternatives in France within the framework of the French National Nutrition and Health Plan – (*Plan National Nutrition Santé*, PNNS).

4.2.1. *What does vegetarianism mean?*

Vegetarianism and its different variations and configurations have been widely identified and described in the literature. The vocabulary used by nutritionists covers a wide variety of definitions depending on the extent of the practice, leading to a

3. FranceAgriMer data, 2018.

4. FAO data, 2009.

complex range of possible categories and self-classification. The richness of the lexicon expresses the diversity of practices around vegetarianism.

It should be noted that these terms have varying degrees of recognition. Some appear in dictionaries, such as "vegetarian" and now "vegan", others are only recognized within communities, especially those online. Similarly, in French, the spelling (and even the pronunciation), under the influence of the English language is, for some words, susceptible to variations (vegan, or veganian, pescatarian). Finally, some terms are specific to the French language, such as "végétalien" (people who do not eat animal products but may buy products such as clothes and cosmetics that include animal products). This word does not exist in English and instead English-speakers usually equate vegans with *either* strict vegetarians (who do not eat meat, fish or other animal products such as eggs) or strict vegans (who not only do not eat any animal products but also reject all kinds of non-food products, such as leather, as well as activities that involve animal exploitation, such as circuses).

Practice	Definition
Flexitarian	Flexitarians follow a vegetarian diet but occasionally eat meat.
Vegetarian	Vegetarians do not eat meat or fish.
Strict vegetarian	Strict vegetarians (sometimes called vegans) do not eat meat, fish, eggs or dairy products.
Ovo-vegetarian	Ovo-vegetarians are strict vegetarians (or strict vegans) who also do eat eggs.
Lacto-vegetarian	Lacto-vegetarians are strict vegetarians (or strict vegans) who still eat dairy products.
Pescatarian	Pescatarians do not eat meat but do eat fish.
Vegan	Vegans refuse to eat or use any animal products or products that use animals (including honey and leather for clothing).
Raw vegans	Raw vegans eat only uncooked and unprocessed foods (fruits, nuts, vegetables).

Table 4.1. *Glossary of practices*

Not eating meat can thus cover a very wide range of realities and of underlying motivations. In practice, as well as in people's motivations and sometimes in their convictions, there are great differences between vegetarians, strict vegetarians and vegans, to mention only the most common classifications. Generally speaking, studies conducted on the profiles of vegetarians show a predominance of female, young and rather educated profiles (85% women in the vegetarian sub-sample and 78% in the omnivore sub-sample, 18% under 30 years old vs. 10% for omnivores, and 40% higher education level vs. 35% for omnivores according to the Nutrinet Santé study (2009–2015) based on a sample of more than 90,000 people (Allès *et al.* (2017)). For all that, we will see that while individuals have a good knowledge of this categorical vocabulary, and spontaneously know how to situate themselves, interpretations and boundaries are shifting. Both individual practices and motivations do not systematically cover the categories. Two main reasons emerge from the previous studies of vegetarians. On the one hand, the variety of motivations, which are often preponderant in the adoption of a given practice, are multiple and often very personal. On the other hand, the practices themselves are part of a long-term life course, which, because of the constraints, contexts and contacts, is evolving over time and subject to hazards. And each individual path has its own history and its own logic.

Beyond the personal journey, the vegetarian must maintain their practice in an omnivorous society, which requires discipline (Greenebaum 2012). This discipline is marked by the vegetarian's "strictness", or the degree to which a person adheres to their specific vegetarian diet (Dagnelie and Mariotti 2017; Rosenfeld and Burrow 2017). The different levels of "strictness" can change depending on the individual and the diet they have decided to adopt. For example, a vegetarian may agree to eat meat at a family meal (Christmas) for social reasons and won't feel that is "cheating" on their practice. An ovo-vegetarian who sometimes eats eggs, in other words, makes deviations regarding animal foods, may feel that they are "cheating". Thus, the levels of observance are eminently subjective and are linked to the very perceptions of identity, the conditions of practice and the justifications for any deviations.

According to the model used by Jabs *et al.* (1998b), meatless diets can be motivated by two main types of reasons: health-oriented reasons or reasons based on ethical concerns. However, for many researchers, classifying vegetarians' motivations into two groups is limited and reduces understanding of the phenomenon and its complex and interactive motivational process (Rosenfeld and Burrow 2017). Following Ruby (2012) and Rosenfeld (2018), four types of vegetarian motivations can be established: health, animal welfare, ecology and religion (however, we will not detail this last point because religion-induced vegetarianism does not relate to a similar process of change. This point will be discussed in another chapter).

The first type of motivation is directly linked to questions of health and well-being. Rather self-focused, these motivations stem from a search for tangible, near-future benefits for the individual's own health (and possibly that of our close circle (family, children)). In a general trend for healthy eating and a growing concern for our health and appearance, consumers are sensitive to nutritional imperatives and recommendations. Meat, often presented as a marker of a too rich diet, is in this respect regularly criticized because of the harmful effects of overconsumption and the risks of becoming overweight, of contracting cardiovascular diseases, cancers or resistance to antibiotics. Other types of motivations are related to global, ethical considerations linked to external reasons, such as concern for the environment, animal welfare or the defense of a responsible consumption and agrifood production model (Beardsworth and Keil 1991; Rozin *et al.* 1997; Jabs *et al.* 1998a, 1998b; Fox and Ward 2008; Hoffman *et al.* 2013). As awareness of the impact that consumption plays in social and environmental issues grows, both at the individual and collective levels, many people are deciding to shift their behaviors (Torti 2017). They have better knowledge of the production mechanisms and marketing chains for meat products and the implications in terms of inputs (water, agricultural land) and negative externalities (pollution, industrialization, loss of local breeds, negative effects on food production or artisanal production). Indeed, these are increasingly described and explained in the media and on social networks. Animal welfare is another important motivation; virulent criticism multiplies, especially on social media, a privileged place for many associations that find there the means to raise awareness with speeches, shock images and emotional as well as informational output (Díaz Carmona 2012; Thomas *et al.* 2019).

This motivational framework is at the heart of the current questioning about the discrepancies observed between attitudes and consumption behavior. While these attitudes constitute a possible explanation for the reduction of meat consumption and the adoption of vegetarian practices, they encounter many resistances and are only translated into behavioral change at the end of a difficult and unevenly shared process. The weight of traditional habits and culture, as well as other hedonic, utilitarian or social motivations often result in the maintenance of consumption behavior. Indeed, meat is considered to be tasty, necessary in a complete diet, noble and indispensable to the conviviality of shared meals. Thus, there is lesser adoption of the vegetarian diet and the practice is still often perceived as counter-cultural and opposed to social norms that are rather favorable to meat. In many cases, sensitivity to environmental and social arguments, though it should favor meat consumption reduction, actually leads to strategies aiming at reducing the psychological discomfort resulting from the opposition between critical attitudes and behaviors that are strongly anchored in terms of social practices and norms. Faced with criticism of meat consumption, consumers first tend to question the importance of these criticisms (decredibilization) or the importance of the stakes and negative consequences of this consumption (trivialization) (Séré de Lanauze and Siadou-Martin 2019). Moreover, meat consumption behaviors are largely defended by social norms and the perception of the difficulties that would be induced by a change in practice. The relational issues surrounding meat consumption are therefore particularly strong, and they considerably influence the process of adopting and maintaining a vegetarian diet.

4.2.2. *Vegetarianism, the steps of a process*

Vegetarianism is based above all on the adoption of a particular diet, which excludes all meat products (vegetarians), or more generally products derived from animal farming and slaughter (strict vegetarians), or even from any form of exploitation of these animals, mammalian or not (vegans). However, beyond the diet alone, vegetarianism is a daily practice that must be adopted in order to ensure this diet, with all its material as well as psychological constraints. The vegetarian diet has an impact on the whole organization of purchase, preparation, taking and sharing of meals, both at home and outside the home, and it is a whole material and social organization that must be considered. Because of these constraints, the convictions that underlie this choice, and the involvement that it requires, vegetarianism also becomes an identity. The vegetarian, while simultaneously defining themselves by their practices (the foods they do not eat), also define themselves by their motivations and the reasons for their commitment. This identity dimension of vegetarianism is then coupled with a social dimension. It is important to understand the levels or dimensions of vegetarianism in order to understand the process of adoption and the evolution of behavior, as well as the relationships and social influences that accompany it.

4.2.2.1. *The steps of the process*

The process of adopting vegetarian behaviors has been widely described in the literature (Jabs *et al.* 1998b; McDonald 2000; Larsson *et al.* 2003; Boyle 2007). It revolves around a key moment, which is the decision to adopt the practice. This decision is the result of two main factors: a set of latent predispositions, and a trigger. The decision to eliminate meat from one's diet is generally rooted in an individual's history:

> And then my family is very "meaty" and very cheesy. I've never been a meat-eater and very cheesy. So already… we were not on the same level (Sonny, study 1).

In particular, certain individual characteristics, certain elements of context and increasing information on the subject prepare the adoption process. Thus, vegetarians questioned about their life choices often report an already weak attraction to the product itself.

It can also be a progressive distancing due to the evolution of the situational context or to a growing awareness of the issues related to meat consumption. During this phase, which can last many years, the individual develops a growing sensitivity to the issue, which leads them to seek and receive information on the subject. Many of the vegetarians interviewed say that they became interested in the practice and its consequences by reading about it in the press or on the Internet. More often than not, they also met vegetarians in their circle of acquaintances who told them about their practices and convictions. But then, it is interesting to note that all vegetarians (or almost all) are able to precisely describe the moment and the conditions of their decision. It usually comes after a specific triggering event, which the individual remembers clearly.

Three types of triggers are most often found, which combine cognitive and emotional components. The first type includes experiences with physical effects or reactions, such as a serious illness, a surgical operation or which generated particularly strong emotions, such as disgust:

> My mother said to me "We're going to see your grandmother, we're going to eat mussels, please don't make a fuss, you will eat them and that's that." And I ate some and then I threw up, and I said to myself "Well, your body made its decision before your mind"… in fact I had forced myself because of the others, because of my mother… and so I said to myself "Well, from this moment on it's my decision, I don't care if it makes the others mad" (Lisa, study 2).

The second type of factor is more related to a particular piece of information, often shocking, often also in the continuity of a growing knowledge of the problem and a quest for more information. In the age of the Internet and social networks, and the growing notoriety and audience of the anti-speciesist movements, many vegetarians link their initial decision to a news report or video they saw on the subject.

In France, the often visual messages of the association L214, which depict particularly shocking animal breeding or slaughtering conditions, and which are relayed on the net in forums and on social networks, are at the origin of the decision of certain individuals:

> I started looking. It's horrible and actually I saw two, I saw three and after a while, but actually, I started crying. I said to myself "It's not possible, it's horrible what they're doing to them" [...]. I said to myself no, but something is happening with your conscience… I said to myself, I'm going to stop eating red meat (Amandine, study 2).

Others are sensitized by exchanges with other vegetarians, who by their credibility (doctor, for example) or their power of persuasion (friends, colleagues) cause the trigger:

> So I was hired and there were vegan people but also non-vegan people who worked there and then that forces a discussion as well, that's it. It fuels the debate and that opens towards other sources as well. People would say to me, "Here, read this book," or "Watch this documentary" and then, there you go, so that's when I became a complete vegetarian (Lucie, study 2).

The adoption of the behavior requires not only a modification of the diet but also the establishment of new habits that break with the past and with the close environment. This involves modifying the content of meals, supplies, recipes and culinary practices, and eating habits outside the home, at work (in the canteen) or in society (invitations, restaurants). The adaptation can be more or less progressive, with possible backtracking or accepted exceptions, especially in the context of commensality.

During the adoption process, the search for information is particularly assiduous. The new practitioner feels the need to find solutions to many material, organizational and relational questions. They need to find new products, recipes, addresses for vegetarian-friendly shops and restaurants and new supplies. Relationally, they must resolve the question of the difference in diet with their entourage and find solutions to ensure compatibility of their diet with the dietary practices of the non-vegetarians they meet or spend time with. Often in the minority in their practice, they are the ones who must often manage situations of conflict or incompatibility. The understanding and flexibility of the people around them, as well as the richness and availability of products and services in their environment, will largely condition the evolution of their own practice. In many cases, a reinforcement of the practice, through increasing demands in terms of diet and the elimination of new foods is observed. The maintenance of the new practice based on a new, well-defined and stable diet is also common. But it also happens that external constraints and pressures are sometimes too strong, and this can lead to backtracking, a relaxation of the rules and an increase in deviations. Here again, relationships with others are key to understanding and explaining the evolution of the practice.

Relational factors are fundamental in understanding the process of adopting vegetarianism. In a cultural context where vegetarianism remains the exception, the vegetarian is often alone in their practice, their entourage (family, long-term friends, colleagues) being little versed in the practice and their environment (offer of products and services) little adapted to their mode of eating. Between the vegetarian and their entourage, relationships can of course take on a variety of configurations. The type of relationship that they establish with others will then largely determine the evolution of the vegetarian's practice, as well as, more broadly, the evolution of their identity and social framework.

4.3. Relationships between vegetarians and non-vegetarians

4.3.1. *From hostility to acceptance*

The vegetarian diet, which is still restrictive in a predominantly meat-based society, generally breaks with previous shared eating habits, with the consequence of generating certain psychological and relational tensions between the people who adopt it and those around them. The social environment, either close or distant, is often not very involved, but its active or reactive role, positive or negative, is a major determinant of the evolution of the vegetarian's practice.

Vegetarianism is the exception in France and most Western countries. In this sense, vegetarians disagree with the vast majority of society. At the same time, the practice is gaining more and more attention from the public and the media and in many respects, is consistent with current concerns. If awareness and a certain acceptance of the practice are growing today, it is no less true that attitudes towards it remain extremely variable and contrasted. Thus, the vegetarian encounters a range of reactions from non-vegetarians (the "omnivores"), which, on a continuum from the most negative to the most favorable, goes from rejection and hostility to acceptance or even adoption of the practice by the other, passing through the stages of criticism, tolerance and interest. In return, each of these reactions from people in the immediate or distant environment will call for attitudinal and behavioral responses from the vegetarian.

The first reaction, hostility, is provoked by a lack of understanding of the motivations, in a rather passive and uninvolved way, as well as and above all by the embarrassment felt because of the constraints and complications generated by the adoption of the practice by the vegetarian whose environment is necessarily shared within the household/workforce. Hostility is then a way of protecting our own behavioral habits, of defending our meat diet and the whole set of culinary and commensal practices that it covers:

> It was more violent on the side of friends, some friends in any case, who said "That's it, you're in…you're boring, it's okay, I won't invite

> you to the barbeque anymore." A whole bunch of stupid things, a lot of very heavy jokes, putting a steak in front of our noses and saying "Oh, that disgusts you?" It's not even funny in fact, but there you go. You get invited and then brought a bowl of rocks, and they tell you, "Here, this is your aperitif" (Lucie, study 2).

The vegetarian's reactions regarding such rejection can go as far as a rupture of the relationship, isolation from the social circle, and in many cases will result in the simultaneous adoption of the diet with departure from the family home. Thus, we frequently observe the adoption of the practice by a student who leaves home to study:

> In general, we cut ties with certain people because they really don't understand (Emma, study 1).

Less significant than hostility, criticism is another form of possible negative reaction from the entourage. It too may be induced by a sense of constraints imposed on everyone by the vegetarian's diet, but more often than not, it covers concerns for the health and well-being of the vegetarian, as often observed in the case of a parent who will try to dissuade their child for fear that they will become anemic:

> His mum was very worried about our health, she was convinced that we were going to be in poor health, very deficient (Lucie, speaking about her boyfriend, study 2).

Faced with criticism, the vegetarian's reactions can be that of contempt or indifference, or can even result in counter-arguments:

> They are not informed, they just see the clichés, and they don't want to hear the arguments. At the beginning I took it very much to heart, today I don't care (Fiona, study 1).

Tolerance is the most neutral form of relational response. Neither positive nor negative, it encourages the same tolerance on the side of vegetarians, because it allows both parties to jointly stand by their positions, both in terms of behaviors and beliefs:

> I don't mind people eating meat, frankly. If everyone was vegan, I'd be happy, but to each their own (Emma, study 1).

More positive reactions can also be observed. The practice can arouse interest in the vegetarian's entourage and lead to a discussion and a willingness on the part of the vegetarian to explain their practice and motivations. Their convictions then feed a pedagogical discourse towards the person who is sensitive to the question and who is asking for information.

Finally, the exemplarity of the behavior can generate the will by one or several members of the entourage to adopt it and the vegetarian will then in response assume a role of accompaniment and sharing of the practice; the examples are not rare within a household, roommates or a work team:

> I met my boyfriend who was not vegetarian at all, but being with me, he started to eat much less meat, and in fact little by little, both of us… (Elena, study 1).

The vegetarian practice is thus subject to many social influences. These are an important driving force in the evolution of the practice and can make it very difficult to eat together with non-vegetarians. They also play a role in reinforcing or destabilizing the vegetarian's convictions. Finally, they often lead to a reconfiguration of interpersonal relationships. Indeed, in order to reinforce their normative comfort, the vegetarian may seek new social relations with fellow non-meat eaters with whom they can often more easily share their meals and experiences with.

4.3.2. *The notion of a vegetarian community*

The question of the group membership arises in many situations where similarities in profiles or practices are subject to the gaze of others who do not share them. The human, as a social animal, feels the need to belong to a group that reassures and protects them, a group that is available to support its members. The group is a source of information, a facilitator of practice and supports the individual's identity. It is also a defense method against the reactions of the social environment outside the group.

4.3.2.1. *Different meanings of the vegetarian community*

When a vegetarian talks about vegetarians, they refer to a community which can cover different meanings, and according to which they will position themselves more or less consciously.

Community can be first of all an abstract entity or community of identity, based on the fact that vegetarianism is a social category. The vegetarian includes themselves in a community of behavioral characteristics as well as of opinions and values. They are thus aware of the existence of a group of other individuals who share certain identity points with them. This group consciousness allows for the reinforcement of our own vegetarian identity, but does not necessarily call for the individualization of group members; the community can remain largely abstract in this respect. Another understanding of community is that it can be seen as a community of practice, a group of people who are more or less identified and gathered around the same practices and beliefs. These communities can be virtual or physical. Such communities can be found on the web where many virtual communities of vegetarian Internet users exchange. But the community can also be understood in a more condensed way, with all the individuals with whom we physically share practices. Finally, groups can have an identity that is both more formal and more collective, refocused around convictions, values or sometimes militant actions, in which case it is a community of convictions.

4.3.2.2. *Expected functions of the community*

The acceptance of a community largely depends on the expectations associated with it in terms of role and function. Several forms of community use are mentioned by vegetarians.

The first of the expected functions is access to information. The community is a major source of information on two levels: information concerning the consequences of meat consumption and non-consumption, from which the perceived benefits and risks of vegetarianism, and information concerning practices are derived. The information helps to increase the individual's awareness and prepares the change of attitude prior to the decision to adopt the new behavior. Excluding meat from one's diet is not a trivial matter. Aspiring vegetarians feel a need to educate themselves on how to do this and the implications for their health. Most vegetarians say that they have found many vegetarian websites to help them develop their knowledge of the subject before changing their behavior. This approach is often initially unidirectional: the individual simply seeks information where it is available, without engaging in discussion or contact with the group. New vegetarians continue after adoption to obtain information through these groups. Information is also necessary to reinforce the validity of their choices and to defend them against attacks and criticism from outside.

A second function expected of the community is that of sharing and exchange. The group is a support to the practice, both utilitarian and hedonic. Exchanges take the form of discussions and meetings, during which recipes, product ideas, addresses of stores, and nutritional and culinary advice are shared. The communities concerned are more tribal in nature, and can be both virtual, such as a discussion forum or a blog on the site of an association or an individual gathering other vegetarians around them, and physical. In this respect, it is interesting to note that the forum of the Association Végétarienne de France (AVF), in its advice on the use of the forum itself, invites interested Internet users to introduce themselves and to specify whether they wish to "possibly get in touch with other people living near them"[5]. Many vegetarians have acquaintances, relatives or friends around them with whom they exchange and share advice with as well as moments of conviviality, which may include meals taken together. Whether they are physical or virtual, these communities are composed of selected, identifiable members with whom dialogue, exchange and sharing are possible.

Third, the community fulfills a social function. It allows the vegetarian to reduce the isolation in which they may feel in a societal context that is still predominantly omnivorous. The community contributes to the reinforcement of the vegetarian identity of the individual and allows them to better resist the negative pressures of social norms from which they have withdrawn. The existence of a community and the possibility of finding a social environment that is consistent with one's choices contributes to the psychological comfort of the individual.

Finally, the fourth and last type of function, the community, fulfills a moral and political role. The latter can be directed towards the promotion of vegetarianism through changes at different levels of public policy. For example, since November 2019, in France, school canteens offer one vegetarian meal per week[6]. Beyond the practice itself, the community contributes to the reinforcement of convictions and the solidification of choices. Vegetarians and vegans frequently mention organizations or associations built around moral, philosophical or political principles that aim to defend the legitimacy of the reasons for the diet and the underlying philosophy of life beyond the practices. In particular, environmental and social issues (defense of animals, fight against waste of resources and pollution, deconsumption and more parsimonious consumption models, terroir and protection of responsible producers, locavorism) motivate more and more individuals to gather within such communities. They become a support for the vegetarian in their quest for meaning, the reinforcement of their convictions and the will to promote a model they believe in.

Table 4.2 summarizes the functions expected by vegetarians in their use of communities.

---

5. See the AVF website: https://forum.vegetarisme.fr.

6. See https://www.20minutes.fr/societe/2640027-20191030-sante-ecologie-economie-pourquoi-essor-menus-vegetariens-cantines-scolaires-bon.

Expected role of the community	Type of community		
	Community of identity *Vegetarian community in the broadest sense and its manifestations in the media and the Internet*	Community of Practice *Selected community of vegetarians known or present on social networks*	Community of conviction *Organized, activist and proselytizing community, associations*
Functional	Information on the issues; development of interest Practice information/ reassurance	Assistance/advice on practices Practice improvement Risk reduction	Defense against out-group attacks
Hedonic	Curiosity/discovery	Sharing practices/ conviviality/pleasure	Participation in debates, actions
Social	Identity reinforcement	Group membership	Social recognition
Political/moral	Awareness	Strengthening of convictions	Support to appeal to non-vegetarians

Table 4.2. *Types and expected roles of communities*

4.3.2.3. *Self-identity and relationship with the community*

Beyond its interests in terms of access to information, advice and sharing, which allow the vegetarian to improve their knowledge, to save time and money (utilitarian benefits) and to find pleasure in a practice that is often perceived as restrictive and frugal (hedonic benefits), the community assumes functions of social and philosophical identity. However, this social role expected of the community varies from one individual to another. In particular, it will depend on our own perception of ourselves as a vegetarian, our desire to identify with the group and our perception of the normative relationships between vegetarians and the rest of society. The issues of identity, social influences and social norms involved in vegetarian community relations are thus particularly central to understanding the influence they have on behavior (Cherry 2015; Vartanian 2015; De Boer *et al.* 2017).

From a theoretical point of view, and especially since the emergence of virtual communities, a lot of research has studied the roles of communities and their influence on individual behavior. Being part of a group is a way to reduce isolation; to reinforce defenses against criticism, aggression and negative perceptions of others and to obtain solutions in everyday behaviors. A social group can be considered a community when group members share common characteristics and values (internalization) and establish relationships based on these values with other group members (Etzioni 1996). These influences can be of two types. Informational influence results from an acceptance of the other's influence based on the perception of congruence of the promoted behavior with the individual's value system. Normative influence expresses the individual's desire to conform to the expectations of the other members of the group for the sake of social identification (Burnkrant and Cousineau 1975). In both cases, the individual defines themselves as a member of the community on the basis of similarities with other members and corresponding dissimilarities with non-members (Ashforth and Mael 1989). From there, identification with a community can have important effects on the nature and evolution of intra- and extra-group attitudes, behaviors and relationships (Dahl 2013; Greenebaum and Dexter 2018). The individual may then, for example, adjust their behaviors to the model of others' behaviors in an effort to associate with them (model effect, Exline *et al.* 2012). However, perceived similarity and sense of community are important moderators. Individuals may prefer to remain distant from other members, in which case the model effect disappears, and even leaves room for the possibility of a reactance effect, if the community model is rejected.

In fact, when they talk about their relationship to a vegetarian community, vegetarians express very different positions. While some claim not to belong to any community, or even refute the very idea of a vegetarian community, others naturally declare a feeling of belonging to a community whose existence they recognize. Belonging to the community then takes various forms. It can be expressed as a simple identity marker, when the individual declares themselves to belong to the community simply because they share the same characteristics:

> If I am part of a community it is not… in fact I know there are other people, but I don't have the feeling… I have the feeling of belonging to a community of people who have decided to eat differently, but they are not close to me (Emmanuelle, study 2).

It can be a real social group but rather limited to the accompaniment and sharing of practice:

> There is no notion of group or clan, it's more like an exchange like this: "You're a vegetarian, what do you eat to make up for the so-called proteins?" (Wilfried, study 2).

It can be a broader ideological community, supporting shared beliefs and values, and reflecting a strong affiliation to a socially identified group:

> I feel privileged in fact because I feel proud and privileged to do things like that, I feel like I am doing something good in any case (Lucie, study 2).

The distinction between network communities and small-group communities here involves personal reasons for seeking out a community (Dholakia *et al.* 2004). Finally, the community can consist of an action group, beyond practices and convictions, in which the individual takes an active part in a desire to affirm and defend vegetarianism and the causes associated with it in society:

> And so I have been involved for several years in animal protection associations, in particular I was close at the beginning, I started to militate with an association for the rights of animals, then with the association FoodA and so, here I am in the animalist party (Fred, study 2).

It is important to keep in mind, however, that the use of these different types of communities is likely to evolve and change significantly over the course of the practice, and in very different ways for different individuals. For example, some people report relying heavily on research on vegetarian websites and then, feeling that their own expertise had outpaced the information available on these sites, abandoned all contact. Others mention the costs in terms of time and effort spent to maintain involvement in exchanges, ongoing discussions and relationships with other vegetarians, which become excessive if the centrality of the topic does not persist in the individual. The particularities of individual diets can also be a restraint to the use of the community from a practical as well as a moral point of view, for example, if the feeling that the judgments made by other more fundamentalist members of the community end up affecting the self-esteem of the moderate vegetarian. Finally, many individuals are uncomfortable with the more confrontational or violent expressions and actions of certain groups. The discomfort in which this places them in their own relations with their entourage is a reason for distancing themselves from the community and raises the question of the community's perceived position within society.

4.4. Opposition between society and community, the normative dissonance

Food consumption is strongly embedded in culture and social norms (Higgs 2015). In France in particular, a country where culinary and table arts are important, the social norm is very strong when it comes to food issues. Here, as in most Western countries, vegetarianism is marginal, not only because of the number of its followers, which is still small, but also because it breaks away from the normative model in force. The vegetarian will therefore have to manage the gap between their practice and this normative model in order to define their place in society. Their perception of the position of the community in relation to society, congruent or opposed, will lead them to resort or not to the community to manage their social identity.

4.4.1. *Conceptualizing forms of normative dissonance (what normative perceptions of vegetarianism?)*

Tajfel *et al.* (1979), the founders of intergroup conflict theory, conceptualize the group as a collection of individuals who perceive themselves as members of the same social category, share an emotional commitment to the same definition of themselves and adopt a certain social consensus regarding the evaluation of the group and their membership (p. 40). Since the connotations, values and therefore social identity associated with a group can be perceived positively or negatively, and since the evaluation of a group is determined in comparison to other specific and considered relevant groups, it follows that a positive identity will be largely based on favorable in-group identifications and advantageous comparisons to other out-groups. Conversely, if social identity is unsatisfactory, the individual will tend either to abandon the group or to make the group more positively differentiated. In either case, the individual seeks a positive social identity that will maintain their self-esteem.

From then on, two fundamental principles make it possible to analyze the social influences of vegetarians with their entourage. On the one hand, the perception of the community as a positive vector of social identity and therefore of social desirability, and on the other hand, the capacity of the group to endorse attributes of social identity that are desired by the individual according to their own self-image. It remains that it is sometimes difficult to determine which is the referent group on which the normative processes will be set up.

In their life stories, vegetarians regularly evoke the cross-views between vegetarians and omnivores in the form of an opposition between two or more groups. They often tend to distinguish, if not to oppose, one group (vegetarians or vegans) with another group (omnivores). Because of their specificities, the first two groups have a high degree of differentiation from the rest of society, whereas the third is practically indistinguishable from it, since it is part of the general normative model. In parentheses, we note that there are also perceived distinctions between minority groups such as vegetarians and vegans, to which we will return later; this differentiation, which can vary greatly from one individual to another, is however less marked than in the comparisons with omnivores.

Group norms must correspond to a positive perception in terms of identity and social norms (Bagozzi and Dholakia 2002). However, vegetarians often spontaneously express the feeling of a more or less effective perceived integration of the vegetarian community into society. Some will see a natural antagonism between the vegetarian community and society, while others will consider that the community is perfectly integrated into society, or even that it is its vanguard and guide. Identification with the community, which is the key to the individual's recourse behaviors and relations with this community, and a product of their social desirability, is also part of a more or less assertive perception of a possible conflict between the group's norms and social norms.

4.4.2. *Perceived normative dissonance between community and society*

The perceived existence and strength of antagonism between the adopted community and the rest of society varies greatly from one individual to another. Some vegetarians feel that vegetarianism is commonly accepted in society and that any difficulties are mainly the manifestation of a marginality that is more numerical than fundamental. They often consider that vegetarianism is integrated in a favorable societal logic, that the tendency is bound to spread naturally in society. Others, on the contrary, point out the frictions between vegetarians and the rest of society, and report structural frictions, rooted in a fundamental cultural and philosophical opposition. They reflect a state of normative dissonance, between the group and the rest of society, which gives rise to psychological discomfort and the need to arbitrate between two radically opposed patterns of norms. We observe that, in a situation of normative dissonance, individuals will favor one or the other normative framework (Figure 4.1).

Figure 4.1 shows how vegetarians are distinguished first according to their level of normative dissonance, but second according to the preferred normative model (society or community). Three broad categories emerge.

The "integrated" vegetarians assume their practice in a natural way and without conflict with the dominant social model. They manage difficulties without cutting themselves off from the rest of society by exploiting all the possibilities of maintaining their practice within the framework of an organization of society, which provides them with the means of adaptation to their environment:

> [Relationships with family are] getting better, but if there's a chicken for Sunday lunch I'll eat it. I'll be more… I won't be confrontational anymore and I'll compromise on that diet, because they've made the meal… so yeah that's one of the… yeah, now I allow myself to eat meat from time to time (Antoine, study 2).

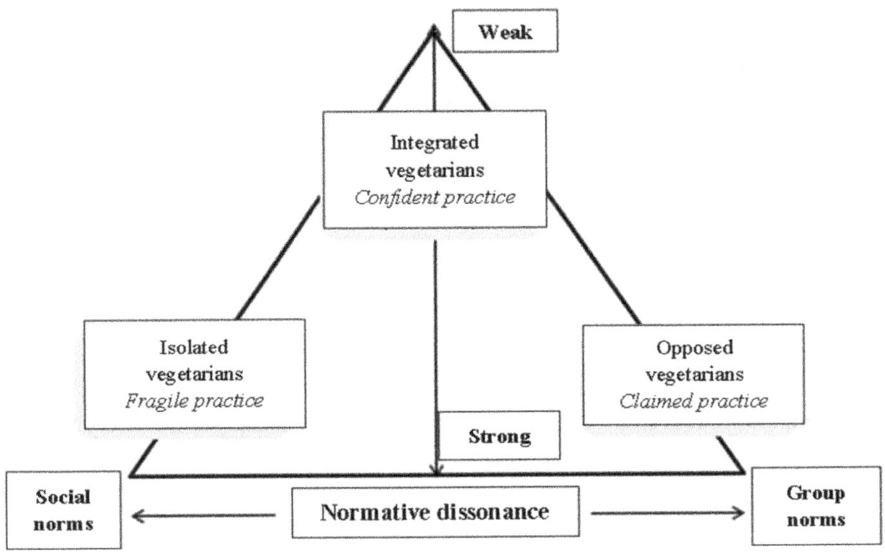

Figure 4.1. *Vegetarians as a function of normative dissonance*

The "opposed" vegetarians maintain their practice at all costs and claim their particularity, in opposition to their entourage. They often adopt militant and proselytizing approaches; they are often found in associations and communities of action, often in connection with other themes that underlie their convictions (environment, de-consumption, defending the rights of animals):

> Then there are people, I know this thanks to my activism experience, you see that there are people you will never be able to convince, so I don't waste my energy on them, you see. But there are people who are questioning, who are open-minded, and even if they are not vegetarians they understand, and I tell myself that my role is to put the little seed in their head (Fred, study 2).

Finally, the third group of "isolated" vegetarians often has difficulty accepting a practice that they believe marginalizes them from the rest of society. Feeling strong pressure to the contrary from those around them, they struggle to defend their choices due to a desire to conform to the dominant normative model:

> I really avoid the subject, whether with family or friends (Fiona, study 1).

It is in this category that we find vegetarians whose practice is fragile, either because it is recent or because the social sacrifices involved are perceived as too high. It is also in this category that deviations from the practice are the most frequent, and at the same time the most resented.

4.4.3. *Strategies for managing and reducing normative dissonance*

Many actors are interested in the emergence of vegetarianism. In particular on the market, food brands, the food trade, out-of-home or collective catering, publishing and trade fairs specialized in nutrition, dietetics, alternative medicine – all are concerned with this trend and are implementing strategies of innovation and promotion of products and services for vegetarians. Other actors, in the markets indirectly concerned, are also trying to understand its evolution in order to anticipate its effects on their own market, in particular of course the meat and meat products market. In their approaches to influencing the market, the arguments in favor of vegetarianism can be based on norms of an injunctive or descriptive nature.

Descriptive norms are based on the perception of what most other people do and are based on the observation of the prevalence of a behavior within the group (Cialdini *et al.* 1990). Injunctive norms reflect the perception of what others who matter to the individual think is right; they express the social pressure of the group's moral rules (White *et al.* 2009). The descriptive or injunctive nature of the normative model used to promote a behavior affects the adoption of that behavior (Smith *et al.* 2012; Stok *et al.* 2014).

Normative arguments are descriptive when they emphasize the exemplarity and social desirability of a behavior shared by the community, based on observed practices and beliefs. For example, arguments are based on more descriptive norms when they emphasize the consensual characteristics and shared benefits of the product and practice for the group. As an illustration, a 2017 advertisement for McDonald's veggie burger highlighting the richness and origin of the product's ingredients (image can be found on the *L'ADN Tendances* website[7]) participates in perceptions of a normal, shared and widespread behavior.

Normative arguments will, conversely, be injunctive when they emphasize the moral position of the referent group with a view to future consequences favorable to the group. For example, the message of a 2019 advertisement for the Hiltl restaurant in Zurich, the oldest vegetarian restaurant in the world according to the Guinness Book of Records, shows a viewfinder whose target turns away from the animal to point at the accompanying vegetable, expressing the moral position of vegetarianism (image can be found on the *Creapills* website[8]).

The effects of such messages will vary across the segments of vegetarians identified earlier. Indeed, the perceived distance to the reference group can moderate the effects of norms on behavioral intentions (Smith *et al.* 2012). Thus, isolated vegetarians will undoubtedly be more sensitive to messages reinforcing the social legitimacy of vegetarianism, through messages based on descriptive norms aiming to bring them closer to a larger number of people, whereas opposed vegetarians will be reinforced in their choices by injunctive reminders defending their convictions, while being less sensitive to demonstrations from the larger number. Finally, unembarrassed vegetarians will probably be the least sensitive to arguments based on standards, and will prefer information with a more directly utilitarian or hedonic content on the attributes and advantages of products and services. Meat product market players have also noticed the importance of the emergence of vegetarianism. Some approaches are interesting in that they also use normative processes in their communication. Thus, the promotion of the flexitarian diet by meat professionals, proposing to consume less but better, is largely based on an injunctive normative process, which tends to confuse the norms of vegetarians with those of the omnivorous society in the same scheme. The approach may be risky in that it relies on a positive model of the vegetarian integrated into society to reach out more effectively to non-vegetarians, while at the same time it could help to reduce the normative dissonance perceived among many vegetarians.

However, these assumptions about the relevance of norms to promote behavior require further research and validation.

The question of the relevance of the evocation and use of the vegetarian community by actors in their innovation and promotion strategies must also be studied in depth. The phenomena of intra-group influences can lead to word of

6. Available at: https://www.ladn.eu/entreprises-innovantes/marques-engagees/mcdo-alors-burger-vege-ou-pas/.

8. Available at: https://creapills.com/hiltl-restaurant-vegetarien-publicite-20191007.

mouth favorable to the products, based on the desirability of the group and the reassurance that it generates, particularly in extra-group relations. How to define and circumscribe these groups and communities and how to give them a voice in order to induce behavior is still largely a work in progress.

Finally, questioning standards also raises the issue of whether non-vegetarians are willing and able to eat vegetarian dishes without being characterized as a vegetarian. A study conducted in the United States among 11,000 omnivores (Krpan and Outsma 2020) shows that the way vegetarian dishes are presented on the menu influences choices. Four presentations were tested: neutral (dishes integrated into the general menu), existence of a "Vegetarian Dishes" menu, an "Environmentally Friendly Main Courses for a Happy Planet" menu and a "Refreshing Main Courses for Relaxing Conversations" menu. The least-chosen menu was the one where vegetarian dishes are presented in a "Vegetarian Menu". Eating vegetarian together, yes, but as long as you can adopt the practice without the identity?

4.5. Conclusion

Vegetarianism is developing in France, with a still limited but growing number of followers, a market that is driven by services (restaurants, catering, shops) and food brands, and an increasing notoriety in the media or on social networks. Research in psycho-sociology, medicine, nutrition and marketing has also largely taken up the subject. The adoption of the practice implies a long and often not very easy learning process and an important domestic adaptation, which are not without effects on the vegetarian's relations with their entourage. It often induces a desire to share with other vegetarians and is coupled with a stake in status and social recognition. The question is thus raised as to *the social enchantment* that such a deserving diet can produce, both from the point of view of the efforts and reflection it requires and the environmental, social and economic issues it addresses. The implications in terms of social influences are thus numerous. The vegetarian is confronted with others within the group of vegetarians as well as outside, with those around them, as well as in their relationship with and within society, which can in both cases constitute a favorable or unfavorable ground. In this respect, normative perceptions are decisive in the process of adoption and evolution of the practice. They contribute to the identification with a model and to the desirability of that model's identity. The role of the group, through the various meanings and manifestations of the vegetarian community, is multiple. As a source of information, as a support for practices, as an addition of hedonic value through sharing, as a reassurance of identity and convictions and as a defense against external aggression, or as a lever for transmitting and imposing values, including through action and proselytizing, the use of the community structures the way in which the vegetarian will find the means to live their practice within society. The community is therefore a central lever for influencing the behaviors of vegetarians as well as potentially of non-vegetarians. Indeed, beyond direct effects such as the "recruitment" of other people convinced by vegetarianism, the influence of the community can also be translated into progress in society, particularly in terms of respect for animal welfare and awareness of the effects of our consumption on the planet.

4.6. References

Allès, B., Baudry, J., Méjean, C., Touvier, M., Péneau, S., Hercberg, S., Kesse-Guyot, E. (2017). Comparison of sociodemographic and nutritional characteristics between self-reported vegetarians, vegans, and meat-eaters from the NutriNet-Santé Study. *Nutrients*, 9(9), 102.

Ashforth, B.E. and Mael, F.A. (1989). Social identity theory and the organization. *Academy of Management Review*, 14, 20–39.

Bagozzi, R.P. and Dholakia, U.M. (2002), Intentional social action in virtual communities. *Journal of Interactive Marketing*, 16(2), 2–21.

Beardsworth, A. and Keil, E. (1991), Vegetarianism, veganism, and meat avoidance: Recent trends and findings. *British Food Journal*, 93(4), 19–24.

Bertaux, D. (2016). *Le récit de vie*, 4th edition. Armand Colin, Paris.

Boyle, J.E. (2007). Becoming vegetarian: An analysis of the vegetarian career using an integrated model of deviance. PhD Thesis, Virginia Tech, Blacksburg, VA.

Burnkrant, R.E. and Cousineau, A. (1975). Informational and normative social influence in buyer behavior. *Journal of Consumer Research*, 2(3), 206–215.

Cherry, E. (2015). I was a teenage vegan: Motivation and maintenance of lifestyle movements. *Sociological Inquiry*, 85(1), 55–74.

Cialdini, R.B., Reno, R.R., Kallgren, C.A. (1990). A focus theory of normative conduct: Recycling the concept of norms to reduce littering in public places. *Journal of Personality and Social Psychology*, 58(6), 1015–1026.

Dagnelie, P.C. and Mariotti, F. (2017). Vegetarian diets: Definitions and pitfalls in interpreting literature on health effects of vegetarianism. In *Vegetarian and Plant-Based Diets in Health and Disease Prevention*, Mariotti, F. (ed.). Academic Press, Cambridge, MA.

Dahl, D. (2013). Social influence and consumer behavior. *Journal of Consumer Research*, 40(2), 3–5.

De Boer, J., Schösler, H., Aiking, H. (2017). Towards a reduced meat diet: Mindset and motivation of young vegetarians, low, medium and high meat-eaters. *Appetite*, 113, 387–397.

Dholakia, U.M., Bagozzi, R.P., Pearo, L.K. (2004). A social influence model of consumer participation in network- and small-group-based virtual communities. *International Journal of Research in Marketing*, 21(3), 241–263.

Díaz Carmona, E. (2012). Perfil del vegano/a activista de liberación animal en España. *Revista Española de Investigaciones Sociológicas (REIS)*, 139(1), 175–187.

Edwards, S. (2013). Living in a minority food culture: A phenomenological investigation of being vegetarian/vegan. *Phenomenology & Practice*, 7(1), 111–125.

Etzioni, A. (1996). The responsive community: A communitarian perspective. *American Sociological Review*, 61(1), 1–11.

Exline, J.J., Zell, A.L., Bratslavsky, E., Hamilton, M., Swenson, A. (2012). People-pleasing through eating: Sociotropy predicts greater eating in response to perceived social pressure. *Journal of Social and Clinical Psychology*, 31(2), 169–193.

Fox, N. and Ward, K. (2008). Health, ethics and environment: A qualitative study of vegetarian motivations. *Appetite*, 50(2–3), 422–429.

Greenebaum, J. (2012). Veganism, identity and the quest for authenticity. *Food, Culture & Society*, 15, 129–144.

Greenebaum, J. and Dexter, B. (2018). Vegan men and hybrid masculinity. *Journal of Gender Studies*, 27(6), 637–648.

Higgs, S. (2015). Social norms and their influence on eating behaviours. *Appetite*, 86, 38–44.

Hoffman, S.R., Stallings, S.F., Bessinger, R.C., Brooks, G.T. (2013). Differences between health and ethical vegetarians. Strength of conviction, nutrition knowledge, dietary restriction, and duration of adherence. *Appetite*, 65, 139–144.

Jabs, J., Devine, C.M., Sobal, J. (1998a). Maintaining vegetarian diets: Personal factors, social networks and environmental resources. *Canadian Journal of Dietetic Practice Research*, 59, 183–189.

Jabs, J., Devine, C.M., Sobal, J. (1998b). Model of the process of adopting vegetarian diets: Health vegetarians and ethical vegetarians. *Journal of Nutrition Education*, 30(4), 196–202.

Krpan, D. and Houtsma, N. (2020). To veg or not to veg? The impact of framing on vegetarian food choice. *Journal of Environmental Psychology*, 67, 104511.

Larsson, C.L., Rönnlund, U., Johansson, G., Dahlgren, L. (2003). Veganism as status passage: The process of becoming a vegan among youths in Sweden. *Appetite*, 41(1), 61–67.

Lepiller, O., Fourat, E., Mognard, E. (2016). Evolutions et facteurs socioculturels de la consommation d'aliments d'origine animale en France et en Europe : état des connaissances. In *Rôles, impacts et services issus des élevages en Europe*, Dumont, B., Dupraz, P. (eds). INRAE, Paris.

McDonald, B. (2000). "Once you know something, you can't not know it": An empirical look at becoming vegan. *Society & Animals*, 8(1), 1–23.

Rosenfeld, D.L. (2018). The psychology of vegetarianism: Recent advances and future directions. *Appetite*, 131, 125–138.

Rosenfeld, D.L. and Burrow, A.L. (2017). Vegetarian on purpose: Understanding the motivations of plant-based dieters. *Appetite*, 116, 456–463.

Rothgerber, H. (2020), Meat-related cognitive dissonance: A conceptual framework for understanding how meat eaters reduce negative arousal from eating animals, *Appetite*, 146, 104511.

Rozin, P., Markwith, M., Stoess, C. (1997). Moralization and becoming a vegetarian: The transformation of preferences into values and the recruitment of disgust. *Psychological Science*, 8(2), 67–73.

Ruby, M.B. (2012). Vegetarianism. A blossoming field of study. *Appetite*, 58, 141–150.

Séré de Lanauze G. and Siadou-Martin B. (2016), Dissonance cognitive et stratégies du consommateur face à une information critique : le cas de la consommation de viande. *Revue française du marketing*, 257, 39–60.

Séré de Lanauze, G. and Siadou-Martin, B. (2019). Dissonant cognitions: From psychological discomfort to behavior change? *Journal of Consumer Marketing*, 36(5), 565–581.

Smith, J.R., Louis, W.R., Terry, D.J., Greenaway, K.H., Clarke, M.R., Cheng, X. (2012). Congruent or conflicted? The impact of injunctive and descriptive norms on environmental intentions. *Journal of Environmental Psychology*, 32(4), 353–361.

Stok, F.M., De Ridder, D.T., De Vet, E., De Wit, J.B. (2014). Don't tell me what I should do, but what others do: The influence of descriptive and injunctive peer norms on fruit consumption in adolescents. *British Journal of Health Psychology*, 19(1), 52–64.

Tajfel, H., Turner, J.C., Austin, W.G., Worchel, S. (1979). *An Integrative Theory of Intergroup Conflict*. Organizational Identity: A Reader, Oxford University Press, Oxford.

Thomas, E.F., Bury, S.M., Louis, W.R., Amiot, C.E., Molenberghs, P., Crane, M.F., Decety, J. (2019). Vegetarian, vegan, activist, radical: Using latent profile analysis to examine different forms of support for animal welfare. *Group Processes & Intergroup Relations*, 22(6), 836–857.

Torti, J.M. (2017). The social and psychological well-being of vegetarians: A focused ethnography. PhD Thesis, University of Alberta, Edmonton.

Tuomela, R. (1995). *The Importance of Us: A Philosophy Study of Basic Social Notions*. Stanford University Press, Palo Alto, CA.

Vartanian, L.R. (2015). Impression management and food intake. Current directions in research. *Appetite*, 86, 74–80.

White, K.M., Smith, J.R., Terry, D.J., Greenslade, J.H., McKimmie, B.M. (2009). Social influence in the theory of planned behaviour: The role of descriptive, injunctive, and in-group norms. *British Journal of Social Psychology*, 48(1), 135–158.

5

Eating Together and Differently: Halal Between Standardization and Segmentation

Foued CHERIET[1], Félix JOURDAN[2] and M'hamed MERDJI[3]
[1] *UMR MoISA, Institut Agro-Montpellier SupAgro, France*
[2] *UMR Innovation, INRAE, Montpellier, France*
[3] *Montpellier Business School, France*

5.1. The halal meat market: eating together or differently?

The term "halal" is an Arabic word that designates what is licit, permitted and legal. Its antonym is "haram", which designates what is illicit, prohibited and illegal. In the Muslim religion, halal and haram can refer to a food product, a practice (life in a couple with marriage) or a financial product (interest and speculation). In the food sphere, the term halal primarily concerns meat products and refers to the way animals are slaughtered. In the last few decades, the halal market has been widely developed, internationalized and standardized, and concerns new categories of food products (confectionery, processed products, soups, baby products and even wine[1]) and non-food products (cosmetics, for example).

1. In France, the company *Petit Béret* has launched a non-alcoholic wine with a halal label.

Evolution of Social Ties Around New Food Practices,
coordinated by Gilles SÉRÉ DE LANAUZE.

Worldwide, the halal food market was estimated in 2018 at US$ 1,400 billion, that is, nearly 20% of food. It concerns national populations and Muslim diasporas, estimated in 2018 to be 1.7 billion people (and a projection of 2.3 billion in 2030). The major markets are those in South East Asia (Malaysia and Indonesia), the Middle East and North Africa. For example, the market in Indonesia is estimated at US$ 200 billion, that of Turkey at US$ 170 billion. Over the last decades, the international trade of halal products has experienced significant developments, encouraging actors of the conventional sector to reorient their production. For example, and only for poultry, the countries of the Middle East and North Africa have imported US$ 5.5 billion (US$ 2.9 billion from Brazil and US$ 400 million from France).

In France, for nearly two decades, halal market industrialists and distributors have become important players in the food system. From an academic point of view, the development of this market raises questions about socio-political[2], legal and regulatory aspects, the organization of value chains (breeding, slaughtering and processing), standardization (accreditation, certification) and finally marketing and distribution issues. Numerous research projects are therefore being carried out to better understand the strategies of the players, consumer motivations and future development trends[3].

The first television advertisement for a halal brand in France (Zakia/Panzani Lasagna) dates back to 2009. During the last decade, this market has experienced a significant boom and rapid growth. In 2018, the halal market in France represented 5.2 billion euros, with 4.5 billion for food, including one billion in out-of-home catering. Of this market, supermarket distribution accounted for only 282 million euros. The halal food market has experienced strong growth (+30% on average between 2010 and 2015) before "stagnating" at 7–9% between 2016 and 2018.

Having reached maturity, the halal market is marked by current trends that resemble what is happening in other food markets: a split between standardization both in terms of production (industrial process and search for economies of scale) and distribution (specialized departments, own brands of major distributors such as Carrefour and large retailers specializing in halal) and search for differentiation: *premiumization* and quality upgrading, niche and differentiated markets (natural halal, debate on animal welfare, quality signs, local halal and specialized Halal Shop).

2. See, for example, the debate between Benkheira and Bergeaud Blackler (2017) published in the *Revue d'Ethnologie Française*.

3. Even though it is important, the level of research in France remains rather limited compared to markets of comparable size.

The development of the halal market in France has been accompanied by a regulatory upgrade, an expansion of offer to conventional players (supermarkets, fast food chains for marketing and expansion of the brands of conventional players to include halal products) and the emergence of new players who develop brands (specialized producers). This development has also been accompanied by numerous political and identity-related polemics, which are often covered by the media and are not subject to rigorous analysis. In addition, there is a "technical" difficulty due to the absence of ethnic statistics in France.

Eating together can be understood through the dynamics of standardization of halal meat on offer (company strategies, slaughter methods, distribution channels, communication, etc.) and homogenization of demand (determinants, purchasing practices, consumption motivations, etc.). *Eating differently* can be appreciated by the efforts made towards hyper-segmentation (search for niche micro-segments) and differentiated meat offer.

In this chapter, the term "standardization" refers to business strategies that consist of offering a product with identical characteristics to several consumer segments (thus meeting homogeneous needs). The terms "segmentation" and "differentiation" are used to describe strategies by which economic operators distinguish themselves from their competitors' offer by a process and/or a product (thus meeting specific needs). The prefix "hyper" is used to emphasize the important segmentation that characterizes a market. The terms used are therefore not intended to reflect possible processes of "social separation", but to characterize the commercial strategies of economic operators positioned in a poorly regulated market.

Beyond the description of these trends of standardization and differentiation that we can observe both in terms of production and consumption, it is a question of understanding the motivations of consumers, the links between these practices and the observed (re)structuring of halal meat supply. The objective of this chapter is to examine the dynamics of differentiation and standardization in the halal meat market. How do these two trends combine within the halal meat sector? What are the main determinants and challenges?

On the demand side, marketing research has pointed out the specificity of halal meat buyers' behavior and the determinants of the purchasing act (religious motivation, belonging, social control, etc.) (Bonne *et al.* 2007). This trend can be likened to "eating differently". For example, the main determinants of overall meat consumption in France are price and quality (especially in terms of how sanitary it is), whereas in the halal meat sector, they refer to a religious practice based on trust and the slaughter method. Similarly, the share of meat purchases in supermarkets (including hard discount) is 77%[4] while it is 14% for halal meat. Purchases of fresh butcher's meat account for 40% of spending on meat products, whereas they are clearly dominant for halal meat. In view of these findings, we think that we are indeed in the presence of a specific market.

On the supply side, the growing development of brands and standardization strategies observed in mass distribution as well as the rise of fast food restaurants reflect a movement of homogenization of the demand for halal meat, recalling the global nature of the movements observed in the food sphere. Moreover, this evolution is accompanied by attempts in the production sphere to reconcile regulatory aspects (supervision of ritual slaughter without stunning, animal protection regulations, meat traceability, etc.) and economic aspects (professionals' desire to position themselves on this particularly dynamic market, process adjustments, search for economies of scale, etc.). Thus, the significant development during the last two decades of the halal meat market has led to important transformations in terms of production systems, logistics, communication and distribution. Finally, certain recent developments in the demand for halal meat reflect new consumer expectations: processed halal products, specialized distribution channels, a demand for labeling and information on the origin and quality of the meat and so on.

The strategic movements of supply actors (production and distribution) and the determinants of demand refer to dynamics of standardization of offer on the one hand and to a search for niches not yet invested in the halal sector on the other hand. This therefore resembles "eating together" strategies. The halal market thus illustrates the problem of eating together AND differently and reflects the coexistence of these two strategic dynamics of standardization-differentiation of offer, responding to an increasingly complex demand. Based on the exploitation of secondary data and those from recent surveys on this sector, we will present the specificities of the halal meat market based on an analysis of slaughtering and processing practices upstream and distribution and consumption downstream.

Assuming a hyper-segmentation of supply, we will address the issues related to the ritual slaughter of animals and the different systems guaranteeing the halal nature of meat. The study of the phenomena of standardization-segmentation will shed light on the economic, regulatory and productive issues underlying the "transformation" of the living animal into a halal meat product identifiable by

4. According to France Agrimer's 2019 studies and synthesis on the food consumption of French households.

consumers. The perceptions of the latter as to the credibility of the certifications or the halal nature of the products will make it possible to make the link between the two levels of analysis (Front End). Finally, we will focus on the demand side by looking at its characteristics, its evolutions, the determinants and the obstacles to halal meat consumption, as well as the way in which demand is articulated with the different sectors available, through the exploitation of the results of a survey and interviews with consumers of halal meat in France.

This chapter on the characterization of the halal meat market in France, both in terms of production methods (supply) and the characteristics of the act of buying and consuming (demand), is based on the collection and exploitation of primary and secondary data conducted between June 2019 and February 2020.

This data collection work was preceded by a broad review of academic literature in economics and sociology of consumption (about 30 recent publications on halal in France and internationally) and professional literature (specialized press, institutional reports, etc.).

The secondary data concern both purchasing and consumption practices, distribution channels, credibility of certifications and brands, and developments during certain religious occasions (month of Ramadan). They come from both reports of the Ministry of Agriculture (2016 and 2019), specialized press files (e.g. LSA articles, from 2015 to 2019), panel data (Nielsen 2019; Ifop 2020) and publications and surveys of the specialized firm Solis (Horizons Shoppers surveys (2010, 2012, 2014, 2018)), as well as various documents from the websites of some production and distribution companies.

The primary data comes from a methodological protocol combining a dozen observations of production conditions in cattle/sheep and poultry slaughterhouses and about 70 interviews conducted with different actors in the sector (conducted in 2018 and 2019 as part of a doctoral project on Muslim ritual slaughter in industrial systems), a series of interviews with consumers (16 people, Montpellier, January 2020, semi-directive face-to-face interviews lasting an average of one hour and 30 minutes) and a questionnaire survey conducted face-to-face with 270 people in Montpellier (January and February 2020).

The main data collected are recorded in Tables 5.2–5.11 in Appendices 1 and 2, sections 5.5.1 and 5.5.2 (description of the survey sample) and in the second part of this chapter. An example of the information from the interviews with the 16 individuals on the purchase and consumption of halal meat is recorded in Appendix 3, section 5.5.3.

Box 5.1. *Methodological approach and data sources*

5.2. Producing together AND differently: actors, complexity and differentiation: segments within the segment

According to various estimates and in the absence of ethnic statistics, 8% of the French population buy from the halal market. According to the Solis firm (2018), 59% of this population only consume halal. The share of meat and charcuterie in this market is 85%, divided for home consumption at 80% between traditional butcher shops and 20% in supermarkets. According to figures from the Ministry of Agriculture (2016), 284 slaughterhouses of slaughter animals practice ritual slaughter without stunning (in very variable proportions from one establishment to another) for a total production volume estimated at 15% of cattle and 27% of sheep (see data recorded in Table 5.10, Appendix 1, section 5.5.1).

The halal meat production chain is quite specific. Respecting halal traditions has little to do with the breeding methods, but there are certain elements that are required during slaughter. These are mainly Muslim religious authorities: three mosques (Paris, Evry, Lyon) that issue sacrificers' cards and halal meat inspection bodies[5] (including AVS, Achahada, Arrissala, Altakwa, etc.).

5.2.1. *Producing differently: actors and organization of halal meat production*

5.2.1.1. *The complex organization of halal slaughter*

In France, the issue of ritual slaughter of animals has historically concerned the production of *kosher* meat, organized since the 19th century around the Israelite Central Consistory. Since then, the definition and control of standards and practices for slaughtering, processing and distribution of kosher meat has remained under the control of the institution. On the one hand, the Consistory has the exclusive right to issue sacrificer's cards to the chohatim (Jewish sacrificers) in charge of carrying out Jewish ritual slaughter in slaughterhouses; on the other hand, it supervises the entire accreditation and control system of kosher products at the different levels of the food chain (Boyer 1998).

The situation is different when it comes to halal meat. From the 1980s onwards, as populations from the Maghreb, Sub-Saharan Africa and Turkey settled (Godard 2007), no central institution played a role equivalent to that of the Consistory for Muslim communities. At the time, the slaughterhouses responded to local orders for

5. OCVH (*Organisme de contrôle de la viande halal*) will refer herein to halal meat inspection bodies.

halal meat, but the most significant volumes were for export markets. Following the example of Socopa International, some slaughterhouses developed a halal slaughter process to respond to public tenders from developing nations such as Iran and Iraq. They are making technical adjustments to enable them to carry out ritual slaughter and receive foreign delegations who come to ensure that practices comply with religious requirements. At the national level, a market of halal retail butcheries is emerging, but it is at this time neglected by professionals who consider it ephemeral and unstable (Bergeaud-Blackler 2005).

The French market for halal meat only really developed in the 1990s. The demand is booming as well as the number of Muslim butcher-retailers, while the global consumption of red meat has been declining for 10 years and export contracts to third countries are ending. Noting the rapid growth of the market, the State tried to regulate and organize it by relying on the Jewish kosher model (Bergeaud-Blackler and Bernard 2010). The government granted the monopoly on the issuance of sacrificer cards to the Grand Mosque of Paris by ministerial decree in 1994, which it finally extended two years later to the mosques of Evry and Lyon. Although no central institution has emerged, the State gives the three mosques the same prerogatives as those granted to the Consistory. Muslim slaughterers are thus able, as is the case for Jewish chohatim (slaughterers), to waive the French obligation to stun animals. However, the lack of centralization leads to the emergence of a market for the control of halal meat. Some OCVH are directly created by citizens of Muslim faith.

With the health crises of 1996–2001, the ensuing restructuring of the beef industry and the decrease in red meat consumption that took hold (Sans and Fontguyon 2003), the halal market became an unavoidable outlet for slaughterhouses. Interministerial surveys conducted in the mid-2000s reported that 46% of large cattle slaughterhouses claimed to perform ritual slaughter (Dunoyer 2008, p. 346), for a slaughter volume approaching 20% of the national tonnage (Riera *et al.* 2005); a figure reduced to 15% in 2014 by the Ministry of Agriculture (Falorni 2016).

For its part, the halal meat control market is developing and evolving. First, the number of national bodies is increasing and the free movement of goods leads to the presence on French soil of halal products certified by foreign bodies. Halal certification has two particularities: there is no unanimously applied specification by the different OCVHs[6] and the State does not recognize or accredit any of them

6. A halal charter was signed by the CFCM (the French Council of the Muslim Faith) in 2016 and by the main OCVHs of the sector, as relayed in this article from the Saphir News website: https://www.saphirnews.com/Charte-halal-l-unanimite-actee-des-associations-

(Tadiello 2017). Several standardization processes are underway at the European and French levels, but no consensus has been established between the different stakeholders (Bergeaud-Blackler and Kokoszka 2017)[7]. This lack of a standard and official certification allows whoever wishes to engage in the business of controlling halal products to do so, and this explains the proliferation of organizations, each with their own certification criteria. Second, OCVHs evolve internally in their accreditation and control procedures. In the 1980s–1990s, companies exporting to the Middle East met fairly simple control criteria. In general, the performance of the slaughter by a Muslim sacrificer and the orientation of the restraint equipment towards Mecca were sufficient. Over time, some organizations have increased their requirements and developed a complex traceability system that includes slaughterhouses, meat processing centers (MPCs), butcher shops and restaurants.

5.2.1.2. *The structuring of the halal meat industry in France*

As we have pointed out, the halal meat market has been characterized since its emergence by the absence of a monopolistic institution holding religious legitimacy and the technical implementation of activities related to halal production. This specificity translates into a diversity of actors, slaughter practices and certification methods, which make the halal meat market a particularly segmented outlet. If there is not one market but several markets for halal meat in France, then there is no halal sector as such. The economic operators within this section are also those who intervene on the other markets; the particularity of the halal market being that it remains largely turned towards the sale in retail butcheries.

To understand how halal[8] meat production is structured in France, it is necessary to place it in the French slaughter and meat processing landscape. Between the 1980s and 2010, the volume of slaughtering provided by public slaughterhouses dropped by more than 60%. While they are still numerous on the territory, service-providing slaughterhouses now contribute less than 10% of the total slaughter volume (Faloni 2016). Most of the tonnage slaughtered is provided by a few large groups, to which are added a number of smaller private slaughterhouses. Halal slaughter is thus "distributed" in a number of these public and private slaughterhouses with heterogeneous practices and halal guarantee systems (HGSs).

---

musulmanes-pressent-le-CFCM_a23178.html. If this charter is supposed to be a point of convergence and a common working basis for the various actors, it has not been the subject of any official statement, each OCVH continuing to apply on the ground its own certification criteria.

7. The latest was published by AFNOR in 2017 under the number XP-V06-001 and cancelled in May 2018.

8. We will focus on the case of cattle slaughter.

On the processing side, some companies specialize in processed halal products (such as Isla Mondial or Isla Délice), while others offer all types of products including a halal range (such as Fleury Michon). Schematically, the market can be represented as follows.

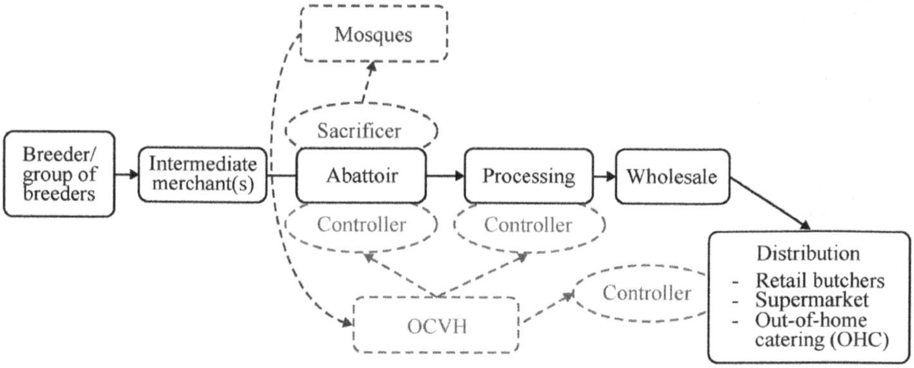

Figure 5.1. *Simplified representation of the meat industry and halal actors. For a color version of this figure, see www.iste.co.uk/seredelanauze/evolution.zip*

In summary, three types of actors are involved: first, the economic operators, second, the mosques that issue sacrificer cards, and third, the OCVHs that guarantee that the production complies with the requirements of their specifications; some of them are directly linked to one of the three mosques. In addition, three types of HGSs currently exist according to the typology proposed by Bergeaud-Blackler (2017), with specificities relating to the slaughterhouse stage.

The first type of guarantee refers to slaughterhouses that operate on a self-qualification (SQ) basis. This approach consists of a company self-certifying itself, that is, to guarantee the halal nature of its products without calling in any OCVH. The second HGS is based on the presence of an audit and compliance agency (ACA) in the slaughterhouse. For this type of OCVH, one-off or initial audits are sufficient to guarantee the halal nature of the products and the production process. Finally, the third guarantee system is the control agencies (CA). The operating mode of these OCVHs is based on the presence, most of the time permanent, of controllers at each halal production and by an important traceability device at all the links of the chain.

Type of HGS	Presence of OCVH	Presence of controllers	Legitimacy of the system	Financial cost of the system
SQ (self-qualification)	No	No	Rests on the sacrificer	Low (no pricing linked to an OCVH)
ACA (audit and compliance agencies)	Yes	Initial, one-off or none	Shared between the sacrificer and OCVH	Intermediary (pricing linked to the right to display the OCVH logo)
CA (control agencies)	Yes	Systematic	Rests on the OCVH	High (fees related to the right to display the OCVH logo and to the control fees)

Table 5.1. *Summary of the three HGSs*

These different HGSs show the complexity of the production and certification of halal meat. In the absence of a central regulatory institution (like the Israelite Consistory), slaughter and control methods vary according to a series of factors (presence or absence of OCVHs, negotiation between management, the veterinary services and the sacrificer(s), clients' requirements, etc.). Thus, sometimes the ways of practicing the Muslim ritual are close to the standards of conventional slaughter (practice of pre- or post-slaughter stunning), sometimes they are far from it (absence of stunning, limitation of contact between halal and non-halal meat, traceability system, etc.). In the first case, the demand for halal meat results in product differentiation (meat stamped or labeled "halal") and little process differentiation; in the second case, it leads to product and process differentiation (to a greater or lesser extent depending on the characteristics of the HGS).

5.2.2. *Producing together: between standardization and differentiation*

In slaughterhouses, the main element of distinction between ritual and conventional slaughter is the use of stunning techniques. These techniques became mandatory in 1964 for slaughter animals and in 1970 for poultry, and were established in slaughterhouses as a standard of animal protection and production in the second half of the 20th century. However, successive animal protection laws provided for an exemption to stunning for religious groups, provided that an appropriate immobilization device was used and that the slaughter was carried out by an authorized sacrificer.

Nevertheless, as there is no binding halal standard, nothing prevents an operator from using a stunning technique if it does not call into question the stability and legitimacy of their HGS. In some cases, we are witnessing a movement of "standardization" of ritual slaughter on the model of conventional slaughter, in other cases a reverse process of differentiation. To illustrate this double trend, we propose to take up the typology of HGS developed previously and to focus on the case of cattle slaughter and then poultry slaughter.

In the case of the *self-qualification* (SQ) guarantee system, two tendencies of standardization and hyper-differentiation are found in similar productive configurations. In the case of small-scale service slaughterhouses, ritual practice can vary greatly from one establishment to another. Depending on the arrangements and power relations between the sacrificer(s), the clients, the slaughterhouse management and the veterinary services, ritual slaughter can be carried out either with stunning (before or after bleeding) or without any form of stunning.

Audit and compliance agencies (ACAs) are generally not prevalent in small service slaughterhouses. They are more prevalent in larger commercial slaughterhouses whose customers require an OCVH stamp to gain credibility. Depending on the size of the company and the extent of its client portfolio, the professional may work with one or more OCVHs – depending on the demands – with different slaughter and certification requirements. In some cases, the OCVH refuses all forms of stunning and imposes measures to limit contact between ritual and non-ritual carcasses; in other cases, the OCVH accepts that stunning is applied immediately after the animals are bled and that halal and non-halal carcasses are mixed together.

The *control agencies* (CA) are particular in that they require the presence of controllers at each halal production. Also, most of them refuse any form of stunning (either before or after the bleeding of animals) and are careful that contact between ritual and non-ritual carcasses is limited to avoid risks of cross-contamination.

These different setups give an overview of the hyper-segmentation that characterizes the production of halal beef. In some systems, ritual slaughter is close to conventional slaughter (we speak of standardization); in others, it is very different (we speak of hyper-differentiation). Between these two opposites, there is a diversity of arrangements according to local contexts and negotiations between the different stakeholders. In general, the more operators deal with a diversity of OCVHs, the more likely they are to split their slaughter process to the point of hosting, sometimes within the same slaughterhouse, several HGSs (where some are opposed and others are similar).

Our investigations have even shown us hybrid forms of HGS mixing self-qualification and control agency. Indeed, some distributors of halal products rely on the presence of a reputable CA in a given slaughterhouse to supply themselves from that same site, without however engaging the services of the OCVH in question. The distributor contacts the slaughterhouse directly and orders carcasses from a slaughterhouse similar to that of the CA. In this way, they switch to self-qualification and base the halal guarantee of their products on the legitimacy of the CA, which they bypass at the same time.

This hyper-differentiation leads to questioning the level of information consumers of halal products have. Does this translate into specific segments to meet distinct demands, expressed by consumers? Do they distinguish between mosques, OCVHs, commercial brands, etc.? In other words, what are the determinants of halal meat purchase and consumption? Are they similar to what is observed in other meat sectors in France? We will try to answer some of these questions in the second part of the chapter, by addressing the question of consumption, the second component of "eating together".

5.3. Consuming together and differently: credibility, trust and differentiation: more segments within segments

Muslim households in France spend an average of nearly 400 euros per month on their food (according to Solis (2018), see Tables in Appendix 1, section 5.5.1) and 54% frequent traditional channels to purchase halal products (halal butchery, grocery stores, specialized stores) and 46% visit halal sections of mass retail. This percentage increases significantly (85%) if we consider only meat/sausage (this sector alone represents 80% of halal purchases). The purchases concern both cut products, ready-made meals and meat-based sauces, soups and baby food.

The purchase of halal meat is highly marked by religious events (Eid, Ramadan, etc.) and reveals certain specificities in consumer behavior. Empirical research conducted on the subject emphasizes the religious nature of halal meat purchase and consumption (apart from the intensity of the person's religious practice) (Bergeaud Blackler and Bonne 2005; Benabdellah 2017). Overall, marketing research, which has mostly mobilized planned behavior approaches, reveals that a positive attitude, perceived behavioral control, habit and tradition, as well as subjective norms and food acculturation, determine halal meat purchase intentions (Bonne *et al.* 2007).

Under the effects of brand launches, structuring of offer and development of specialized departments in supermarkets, the trends here also oscillate between standardization of consumption (products, brands, places, purchasing practices and communication) and more differentiation and segmentation of this already specific market.

While halal guarantee systems raise issues in terms of production methods, they also have an effect on consumption behavior. For example, our questionnaire survey (see results in Tables 5.7–5.10 in section 5.3.2) revealed that the halal nature of meat was essentially linked to its method of slaughter (for 94% of respondents), whereas we are indeed in the presence of "goods of faith".

Below, we examine the main specificities of halal meat consumption. Some trends (standardization vs. differentiation) are consistent with what is generally observed in food supply and demand practices; others are more specific to the halal meat market.

5.3.1. *Eating differently: specificities of the halal meat market in France*

Over the last two decades, the halal market has undergone many changes, both in terms of size and players present and in terms of consumer practices. The expansion of offer was first done by the emergence of medium-sized companies developing national brands (Isla Délice, Isla Mondial, Reghalal, etc.) and the expansion of the range of large food companies (Zakia Halal de Panzani, Maggi, Herta, Fleury Michon, Labeyrie, etc.) to meet the demand of traditional channels. It has continued with the gradual introduction of halal products on the shelves of supermarkets, the launch of specialized outlets (Halal Shop), the development of a specific offer in franchised restaurants (launch since 2009 of a halal offer at the fast food chain Quick, present in 10% of the chain) and finally the launch of generalist distributor brands (Carrefour Halal, Wassila for Casino, for example).

This restructuring of offer was both a response to and a stimulator of the evolution of demand for the halal meat sector. The growth rates of demand between 2010 and 2015 were around 30% with about 7 to 9% stabilization from 2016. In large retailers (LRs), which remain a secondary sector of halal supply, especially for meat, the market for halal products estimated in 2009 at 80 million euros reached 280 million euros according to estimates by Nielsen (2019). The halal sector represents only 0.3% of the total product offer in LRs; however, the average annual growth rates of around 5% have led many brands to propose a specific offer, to restructure their shelves and to develop, with varying degrees of success, their own halal brands.

In supermarkets, there is both a concentration of purchases made by large buyers (76% of the total) and a concentration of sales mainly made in hypermarkets (65% of the total). Thus, according to the Nielsen (2019) panels, 4% of outlets achieve 50% of halal sales, and 43% of stores reach 98% of sales of halal products, all categories combined. Sales in LRs mainly concern fresh meat, charcuterie and meat products (frozen or prepared), representing 90% of the total.

In terms of profiles, the buyers of halal products in LRs have family profiles (46% have children under 11 years of age) and are young (77% are less than 50 years old). Purchases of these products increase significantly during the month of Ramadan (+11%) encouraged by specific promotions (+42% against 20% on average). The purchases record strong increases during the first two weeks of this month. Regarding the perception of halal offers in LRs, the Solis surveys (2014 and 2018) show a notable improvement both in terms of shelf organization, offer and the diversity of brands offered. Nearly 70% of shoppers show a clear preference for halal offers to be located on the same shelf.

Regarding consumption habits and consumer perceptions, several elements accentuate the specificity of this segment: first, the systematic consumption of halal products is high. According to a 2020 Ifop survey, it reaches 67% of French Muslim households (and 79% for practicing Muslims). Our consumer survey revealed an average proportion of systematic consumption of 76%. Other surveys found comparable results. Bergeaud-Blackler and Bonne (2005) conducted a survey of 576 people in France. Their results indicated that 85.9% of the respondents reported systematic consumption.

When it comes to meat, it is important for Muslim consumers that they are eating halal. According to Solis surveys (2018), 86% of households that reported only buying halal products, are sensitive to the halal nature of products when it comes to meat against 24% for ready-made meals, 22% for soups and less than 10% for baby food.

In addition, under the effects of standardization of the communication of food companies (specialized or generalist) and supermarket chains, shoppers say they know more halal brands on average (8.7 in 2018 vs. 6.8 in 2014). Finally, according to the Solis Horizons Shoppers surveys, in 2018, 44% of consumers make their purchases whilst taking into account the guarantee that what they are buying is halal (vs. 36% in 2014) and 72% would and can cite at least three halal certification bodies (vs. 61% in 2014).

These initial factual results help to account for the specificity of the halal meat market in terms of consumer behavior. Previous studies have reported many motivations beyond the religious character (Bonne *et al.*, 2007; Benabdallah 2017). Thus, while consumption is a religious act, some consumers argue that this meat is also safer in terms of health, is more hygienic, is of better quality, has a better taste, and is from a mode of slaughter that is less painful for the animal (Bergeaud-Blackler and Bonne 2005).

These motivations are moreover observed in other European countries: in Germany for the community of Turkish origin (Sherwani *et al.* 2018), in Belgium for the community of North African origin (Bonne *et al.* 2008), in Scotland for the Middle Eastern community (Elseidi 2018) and in Austria for the Turkish community and from Eastern European countries (Meixner *et al.* 2018). All of these studies, which share the mobilization of the theory of planned behavior, account for the religiosity of the act, the community belonging effect, and the halal "label" as determinants of the purchasing act.

These specificities of the act of consumption have not prevented a response through the development of a standardized offer by mass retail and large agrifood companies, fast food restaurants, an "industrialization-financialization and internationalization" of the traditional circuit itself: dedicated brands, international "halal" financing chains of butcher shops and halal outlets, halal shelves in LRs and private labels, expansion of the range of food groups, standardization of promotional practices, communication and merchandising and so on.

This movement towards standard halal is itself creating resistance among certain actors: certification bodies would like to retain control over halal standardization, other actors are specializing while some "hyper-segment" by advocating "authentic halal" to counter the "merchant halal" sector.

5.3.2. *Eating together: between standardization and hyper-differentiation*

The exploitation of secondary data from institutional reports, specialized press reviews and surveys of ethnic marketing consulting firms (in particular, the Solis firm) has shown the coexistence of the two trends of standardization and differentiation, both in terms of distribution channels, brand strategies, product lines, halal stamps, communication and promotion tools, and halal meat purchase and consumption behaviors.

Furthermore, the results obtained from previous empirical research conducted in Europe (based on surveys or via the collection of factual data) all confirm certain specificities of purchase determinants and consumption behaviors, especially in the halal meat sector.

In the following, we will attempt to confirm the main trends observed based on the results of our primary data collection. As a reminder, the latter is based on a series of semi-structured face-to-face interviews (16 people in November 2019) and a face-to-face questionnaire survey (270 respondents in Montpellier, January–February 2020) (see Box 5.1, found in the introduction to this chapter, and Appendices 2 and 3, sections 5.5.2 and 5.5.3).

Regarding the quantitative survey sample, the age and gender distribution of respondents was homogeneous for a survey of this nature. 38% of respondents were under 30 years of age, 55% between 30 and 60 years of age, and 7% over 60 years of age. Men were slightly over-represented (57%) compared to women.

The majority of respondents were French (71%). The other nationalities represented (29%) were from the Maghreb countries (Algeria, Morocco, Tunisia), some Sub-Saharan African countries (Mali and Senegal) and three Middle Eastern countries (Turkey, Syria, Lebanon). The respondents were born in France (47%), in the Maghreb countries (45%) or in other countries (8%). Finally, their parents had the nationality of one of the three Maghreb countries (80%), were French (13%) or had other origins (7%). They were therefore essentially first- or second-generation immigrants, or spouses of people of Maghrebi origin or of Muslim culture. Respondents were approached at the exit of halal butcher shops, in halal departments of LRs and in the "popular" neighborhoods of Montpellier: Plan Cabane, Figuerolles, La Paillade, Les Cévennes, Les Hauts de Massane, Celleneuve, etc. The results concerning purchasing halal products (more specifically meat) and consumption behavior are recorded in Tables 5.2 and 5.3.

Meat occupied a predominant place in the purchases of halal products (98.5% of responses against 28.5% for ready-made meals), with an average frequency of purchase of once a week (40%) for an average budget of less than 50 euros (64%).

In terms of channels, halal butcher shops accounted for 87% of purchases against 20.7% for traditional grocery stores, while supermarket chains monopolized 44% of purchases. The latter two accounted for purchases of ready-made meals, cold cuts, and frozen products and meat derivatives, while traditional butcher's shops were more concerned with cut meat (white meat and poultry for 82% and fresh red meat for 77% of respondents), and to a lesser degree frozen meat products.

The certainty that the product was halal was not the first determinant when choosing the place of purchase. Proximity to home, hygiene and quality of products were the first three determinants (with respectively 68%, 60% and 54%) and the certainty that it was halal only came in 4th position for nearly 41% of the answers.

Items	Frequency in % of cases
Halal products consumed in general (the first three answers)	
Meat	98.5%
Charcuterie	45.20%
Ready-made dishes	28.20%
Places of purchase of halal meat products (the first three answers)	
Halal butchers	87.8%
Super and hypermarkets	44.1%
Grocery stores	20.7%
Criteria for choosing where to buy meat (top five answers)	
Proximity to home	68.2%
Hygiene	60%
Product quality	54.1%
Certainty as to the halal nature of the products	40.7%
Contact with the seller	33.7%
Frequency of purchasing halal meat (the first three answers)	
Two to three times a week	26.3%
One time per week	39.6%
Two times a month	18.2%
Average monthly budget for halal meat purchases (first 3 answers)	
Less than 50 euros	64.1%
From 50 to 100 euros	21.1%
From 100 to 150 euros	7.8%

Table 5.2. *Halal products and meat purchase behavior (source: elaborated by the authors – Halal Meat Survey, Montpellier, January 2020)*

In the traditional channel (halal butcher shops and grocery stores), the offer of meat and derived products is considered sufficient and of good quality. In terms of brands, the offer is perceived as more complete than in the halal departments of LRs, including for processed products (cold cuts, frozen foods, ready-made meals). The diversity of what's on offer can determine the choice of the place of purchase, as many respondents indicated that their practice is not limited to meat but also covers dry goods, other animal products (eggs, etc.), traditional bread, condiments and so on.

For the majority of respondents, the consumption of halal meat goes beyond a simple choice in terms of food purchasing behavior. It is linked to a religious practice for a large majority of respondents (87%) and the purchase of halal meat is systematic for three quarters of respondents. As a reminder, this result was 59% (and 91% for practitioners) in the Solis survey (2018). It is only occasional or rare for 24% (28% in the Solis 2018 survey). This was confirmed to us by the interviews[9]:

> The consumption of halal meat is for me a religious act, it is in my traditions. It is the Muslim religion that requires it, it is strictly not a pressure but rather a choice on my part to continue to eat halal (Lina, 18 years old, single, of Turkish origin).

> It's a religious issue [...], it's not for tradition, it's just for religion (Madjid, 26, single, French-Moroccan).

For other consumers, this purchase is more of a habit:

> I eat halal meat because of "tradition" because my wife is Muslim. Today I eat it out of habit (Pascal, 64 years old, French, married to a Moroccan woman).

As already noted by the Solis surveys (2018), this consumption increased significantly around religious events, especially the feast of the sacrifice of the sheep (Eid al-Adha, 73%) and during the month of Ramadan (70%). Other religious events (Eid al-Fitr (end of Ramadan), Mawlid an-Nabawi, Achoura, etc.) or family ceremonies (weddings, engagements, circumcisions, etc.) were also the occasion of increased demand, for 40% of respondents.

These initial results are in line with those obtained by previous research. Bergeaud-Blackler and Bonne (2005) showed that if the first motivation for the

9. Collected and translated from French by the authors.

purchase and consumption of halal meat is religious, other determinants such as health, hygiene and quality are common concerns.

Thus, in a work based on 45 interviews with people from three generations of immigrants, Benabdallah (2017) shows that the majority of interviewees (and more for those from the first generation) maintained that halal meat is safer (slitting the throat reduced the risk of contamination), tastes better, and that beyond the religious practice, purchase is perceived as an act of community belonging.

Items	Frequency in % of cases
Consumption of halal meat, a religious practice	
Yes	86.7%
No	13.3%
Systematic nature of consumption	
Systematically	76.3%
Occasionally	19.3%
Rarely	4.4%
Events that promote a significant increase in the consumption of halal meat (the four main answers)	
Feast of Eid El Kebir (Sacrifice)	73.3%
Month of Ramadan	70%
Family ceremonies (wedding, circumcision)	41.9%
Other religious holidays (end of Ramadhan, Achoura, El Mawlid, etc.)	39.3%
Main halal meat products purchased (top three responses)	
Fresh white meat (whole chicken, drumstick, wing, leg, etc.)	82.2%
Fresh red meat (beef, veal, lamb)	77.4%
Merguez sausage	14.8%

Table 5.3. *Halal meat consumption behaviors (source: elaborated by the authors – Viande Halal survey, Montpellier – January 2020)*

The reluctance to buy halal meat from large retailers may be linked to lack of confidence in the halal nature of products, as shown in some interview extracts:

> We do not consume halal meat that is sold in supermarkets, even though it is well presented and clean. We are not used to buying pre-packaged meat, and halal meat is usually found next to meat that is not halal and even next to pork. These are the two factors that make us not consume halal meat from supermarkets (Lina, 18 years old, single, of Turkish origin).

These barriers are also related to issues of perceived quality, trust and the freshness of products:

> In supermarkets, the meat is wrapped in film. It is possible to change the expiration date, but this is something I don't think I will find in the butcher's shop where I go. I see the product, if I buy minced meat, [and then] they chop it in front of me (Mohamed, 56 years old, married, French-Moroccan).

> I now enjoy eating halal meat because it is a quality meat, a very fresh meat. Having previously bought non-halal meat in supermarkets, I notice the difference. And for nothing in the world will I eat non-halal meat again (Pascal, 64 years old, French, married to a Moroccan woman).

These obstacles when purchasing halal meat in supermarkets can be explained by the lack of separation between halal and non-halal meat departments, mistrust of the halal nature of products in LRs, a perception of better value for money at the halal butcher's, difficulty in identifying halal offers on the shelves or a feeling of stigmatization by other consumers in the store (Benabdallah 2017).

Concerning the halal nature of a product and the information allowing a potential buyer to be sure of this, the results obtained show the significant part played by proximity and confidence granted by the butcher compared to knowledge of the process and certification organizations, to the relevance of the information conveyed by the labels or signage on the shelves or in the store and even less to the notoriety of the brands. Thus, while the information contained in the labels of halal products is considered sufficient (for 47% of respondents), less than 20% look at these labels at the time of purchase and only 57% check if the label mentions the word "halal".

For 86% of the respondents, the halal nature is based on trust in the butcher, against only 24% in a certification body and 17% in some brands. One of the respondents reinforced this finding about trust in the butcher:

> I always go to the same butcher's shop, because there is this relationship of trust between the owner and me. The owner has become a friend of mine (Aziz, 32 years old, married, French-Algerian).

The signage on the shelf or in the store completes the butcher's trustworthiness. 36% of respondents attach importance to this, but without requiring a particular label or certification:

> The word halal and the certifications are displayed at the butcher's. This proves and reassures us that the meat we buy is indeed halal. The medium used [to communicate this] are paper posters (Lina, 18 years old, single, of Turkish origin).

Items	Frequency in % of cases
How to verify halal (sic main answers)	
I trust the butcher	85.9
I check the existence of the word "halal" on the product	56.7
I check the signage of a halal department or store	35.6
I trust one or more certification bodies	23.7
I look at the information on the labels	19.6
I trust some brands	16.7
Believing that there is sufficient information on product labels about whether or not the product is halal	
Yes	47.00
No	22.60
No opinion	30.40
What is the meaning of halal meat (top five responses)?	
The method of slaughter of the animals	93.7
The quality of the meat	21.5
The way the animals are raised	13
The way the meat is processed	8.5
The place of distribution of the meat	4.1

Table 5.4. *Significance of halal and selection of place of purchase (source: elaborated by the authors – Halal Meat Survey, Montpellier – January 2020)*

Other respondents confirmed this role of signage, but suggested other subjective elements: the area where the butchery is located, the North African origin of the butcher, the presence of writing in Arabic and so on:

> For verification of halal meat I rely on the front used at the butcher's, where it is written "halal" in French, then there is writing in Arabic. Of course, I cannot read Arabic, but it brings a reassuring element. We tell ourselves that the meat we will find inside is halal, moreover, [when we see] the paper that is inside the store where we see the halal certification. And finally, the relationship of trust with the butcher, it is a North African person, and for me he would not sell non-halal products (Fatou, 38 years old, married, French of Malian origin).

These results are supported by other empirical research. Thus, Benabdallah (2017) has shown that the butcher benefits from confidence in the halal nature of the meat, from a better perception of the quality of the products and lower prices, as well as geographical and cultural proximity. The butcher is also perceived as a community representative and a relay of ethnic and religious information. Finally, the butcher's shop is seen as a place of social gathering and of supply of other ethnic products.

In the same study, differences were observed between people of the three generations. While those of the first generation preferred butchers of the same nationalities, those of the second generation had a preference simply for butcher shops in general, without distinction of nationalities, and those of the third generation indicated an equivalent frequentation between supermarkets and halal butcher shops, with determinants of choice related more to price and location.

While the Solis surveys (2018) reported good unaided awareness of certification bodies (72% of respondents in 2018 were aware of at least three certification bodies, compared to 61% in 2014), our survey found that only 22% of respondents were aware of these bodies. Of those, only 21 respondents talked of the credibility of the organizations and the certification process.

Among the organizations considered credible, we find the three mosques (Paris, Evry and Lyon) and others. Moreover, 94% of respondents perceive meat as being halal depending on the method of slaughter. Very few took the breeding (13%), processing of the meat (8.5%) or the place of sale (4%) into account. This last point raises questions because buyers also declare that they consider the place of sale (or even the neighborhood) when looking at options. They attach great importance to the butcher's shop and to the information present at the point of sale.

A wide range of responses on the meaning of halal, the method of slaughter and the credibility of the certifications were noted in our interviews. Overall, respondents referred to "artisanal" slaughter and noted that ritual slaughter must be done by a Muslim, in the direction of Mecca and according to Islamic precepts. No respondents reported an industrial mode of slaughter. Thus:

> The meaning of halal for me is in the method of slaughter. To kill the animal in the direction of Mecca, and to say "bismillah" before slitting the throat. I don't see any other meaning of halal (Olfa, married, 46 years old, French-Syrian).

Unsolicited knowledge of certification bodies	
Yes	22.2%
No	77.8%
Certification bodies deemed credible (based on 21 responses)	
Mosques of Paris, Evry, and Lyon, Altakwa, Oriental Viande, Achahada	
Brands with high halal awareness (from rank 1 to 6 out of 10)	
(1) Isla Délice, (2) Maggi, (3) Fleury Michon, (4) Carrefour Halal, (5) Reghalal, (6) Isla Mondial	
Top six brands in terms of trust/halal nature	
(1) Isla Délice, (2) Reghalal, (3) Isla Mondial, (4) Carrefour Halal, (5) Fleury Michon, (6) Zakia Halal	

Table 5.5. *Halal certification and brand awareness (source: elaborated by the authors – Halal Meat Survey, Montpellier, January 2020)*

Concerning the credibility of the certifications, the religious nature of the act of purchase and consumption gives the three mosques a certain legitimacy:

> The certifications come from the three major mosques, which is a guarantee for us. This provides confidence (Lina, 18 years old, single, of Turkish origin).

Other respondents questioned the control capabilities and certifying role of mosques for a food product:

> For the credibility of organizations, I do not find it especially normal that a mosque is able to deliver a certification on such a product. It is not nothing to certify that a product is halal. And I do not see how they are able to certify it (Pascal, 64 years old, French, married to a Moroccan woman).

Finally, brands still seem to be marginal in terms of purchase determinants in traditional channels (butcher shops and grocery stores). Questionnaire surveys and interviews show that the influence of brands is rather weak and that consumers have an average knowledge of the brands associated with meat products (54% of respondents mention at least four brands).

Contrarily, in the LR circuit and concerning processed products (frozen food, ready-made meals, meat-based products), our survey showed a certain balance in terms of awareness and trust between specialized halal brands (Zakia Halal, Isla Mondial, Reghalal), those belonging to large agrifood groups (Maggi, Fleury Michon) and private labels (Carrefour Halal). The Isla Délice brand is widely acclaimed by consumers, ranking first in terms of both awareness and trust.

This rise of halal brands, via specialized players or by expanding the ranges of large food companies, is not without risk for the credibility of products and consumer confidence. Wilson and Liu (2010) pointed out in this sense that there was a certain cognitive dissonance and lack of understanding with regard to brands with a strong reputation displaying a halal label, but that the latter remained a source of differentiation and gains in competitiveness. These authors had moreover noted this tendency to hyper-differentiation, both in terms of product offers and practices of large retailers and food groups.

Overall, and despite efforts to standardize, this sector remains specific due to the nature of the act of purchase (religious), its seasonality and dependence on religious events, and the trust placed by consumers in traditional players, keeping the halal butchery circuit popular. Certain constraints upstream (economies of scale, homogenization of the industrial process and adaptation of the slaughtering method) and downstream (decrease in demand) may arbitrate future trends in the halal market in France.

Generally speaking, future developments in the halal meat sector of France will be characterized both by differentiated consumption/standardized production trends (trend 1, the most dominant case) and increased searches for *standard* consumption models requiring industrial optimization and an alignment of company strategies (in terms of distribution, communication and brands, characterizing trend 2 with the efforts of LRs). Two other trends seem to be emerging: the first is related to attempts to "standardize" consumption but with a differentiated offer strategy. This would be the prelude to the birth of a specialized supermarket and hypermarket as well as exclusively halal brands (trend 3). Even though some examples are recorded here and there, this trend would be viable only through considerable efforts in terms of differentiation of the modes of production, something that the industrialists could not agree on in view of the market being so limited in size. Finally, a last trend is worth mentioning. This is a market configuration with both differentiated offers and demand. This would be compatible with niche sectors and increased demand for specific products. In this case, small production players could invest to differentiate their offer and meet a demand for specific attributes of halal meat.

It thus appears that the problem of "eating together AND differently" applied to the halal meat sector in France takes into account both the determinants of production processes and the strategies of offer actors, as well as the specificities of demand and expectations of these consumers.

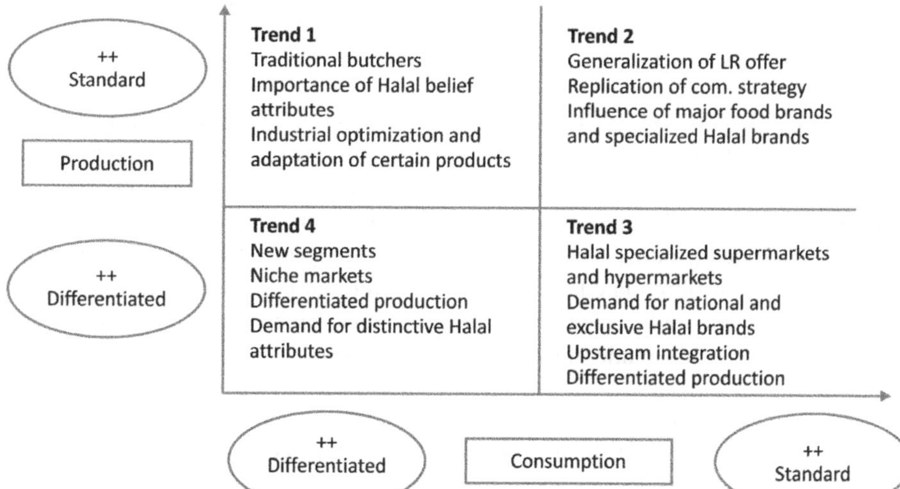

Figure 5.2. *Eating together, standardization and differentiation trends in the halal meat market*

5.4. Conclusion: the halal meat market in France: eating together and differently

In this chapter, we have looked at the concepts of "eating together" and "eating differently" in the halal meat market through the dynamics of standardization versus differentiation, both in terms of purchasing practices and offers, as well as in terms of meat production methods and guarantees of their halal nature. Two concomitant trends are observed: on the one hand, a standardization of offers, encouraged by the involvement of LRs and food groups which widen their range. On the other hand, there is an increasing differentiation between traditional distribution players, industrial companies and religious intermediaries, some of whom wish to escape the standardized halal market and retain control of historical niches and "authentic" halal practices (Lever and Miele 2012).

Eating halal meat is a religious practice for the majority of French people of Muslim origin and/or faith. For them, it is a question of *eating differently*. This sector is distinguished by its consumption motivations and its seasonality, its actors and institutions involved as well as the organization of the downstream industry. The development of halal meat that is available has affected all the distribution channels and food actors, even though the historical actors and traditional channels remain dominant. This movement of standardization is thus similar to the dynamics of *eating together*.

While standardization practices are more common in LRs and chain restaurants (harmonization of offer, expansion of product ranges, reorganization of shelf space, adaptation of promotional strategies to the seasonality of consumption, etc.), the traditional channel is also making efforts in this direction: increasing the variety and number of branded products on offer, integrating ready-made meals, seeking niche sectors through a differentiated and innovative offer that is generally in line with the standard offer observed elsewhere (foie gras, duck breast and halal charcuterie, organic products, ready-made meals, by-products, etc.).

This coexistence of the two trends of standardization-differentiation to varying degrees in the two main halal products marketing channels depends on the adaptation of the supply of halal meat in terms of slaughter and industrial production. This is in line with the issues explained in the first part of this chapter: which halal guarantee systems? What is the credibility of the actors and the certification systems? What adaptations of the industrial production tool?

This sector differentiated both upstream and downstream is also specific because it responds to a *demand for difference*. Eating together through practices, places of purchase, brands (halal range in LRs, or at Quick, Fleury Michon Halal) and eating differently also through other practices and in other places of purchase (traditional channels, halal brands). As such, the players in this sector are all trying to influence the matrix of all possible niches: from standard with private labels in supermarkets to natural halal in specialty stores. It is conceivable to imagine "Islamically responsible" halal circuits in short circuits or an alterglobalist halal as a response to the generic and standard halal of a globalized market (Bergeaud-Blackler 2017).

This search for differentiation would probably make the halal market resemble other specific markets (organic, fair trade), which have globally experienced the same standardization-differentiation trends, with the difference that the halal market, beyond the market sphere, is at the crossroads of politics, religion and various societal issues.

In addition, it should be noted that this sector, although specific, is also very uncertain, for various reasons that relate both to the nature of the offer and its actors, as well as the evolution of demand and regulatory institutions (particularly in terms of credibility and organization). There has been a sharp slowdown in the growth rates of halal sectors, particularly in supermarkets, after nearly a decade of strong growth. This may mark a certain resistance to standard halal and the beginning of the development of an alternative and niche halal.

Then, it is important to note that the future evolution of the halal meat market in France strongly depends on the strategies of the actors involved: on the one hand, on the adaptation of the production tool and its capacity to achieve economies of scale (or even to find sources of funding halal), on the one hand, and the innovation capacities of supply actors in response to the evolution of consumer demand and societal issues (halal in short circuits), responsible halal (in the sense of CSR (consumer social responsibility)), processed products, signaling of quality, specialized halal point of sale in LR format, Islamically responsible halal (financing, operation, production, etc.), on the other hand.

Finally, whatever the future dominant trajectory of this market, the two "standard and alternative" models will not escape increasing efforts to differentiate (or even hyper-differentiate for niche markets) the need to coexist and the need to resolve the many societal controversies related to this sector (halal guarantee, animal welfare, influence of religious bodies, etc.). Eating together AND differently should continue to characterize the halal meat market in France in the years to come.

5.5. Appendices

5.5.1. *Appendix 1: Secondary data on the halal meat market in France*

Characteristics/ sources	Data
Market size and growth (Solis 2018)	5.5 billion euros, including 1 billion in the catering industry (and 4.5 in food products); a 30% increase between 2010 and 2015 and 7 to 9% per year between 2016 and 2018
Retail figures (Nielsen 2019)	+5.2% annual average 80 million in 2009, 170 million in 2013 and 282 million in 2019 Slower growth: +20% two years ago, +5% in 2019 Halal = 0.3% of consumer products in total LRs (traditional channels still very present) (this increases to 1% in 5 stores)
Ramadan figures 2019 (Nielsen 2019)	+11% of sales 42% of halal sales in promotions (vs. 20% on average) 32 million euros in five weeks, that is, +10%/2018 +19% compared to an average week +45% the previous week and the first week of Ramadan
Halal LR characteristics (Nielsen 2019)	Sales: hypermarkets: 65% and supermarkets: 15% In hyper and supermarkets, 4% of the outlets achieve 50% of turnover 43% of the stores make 98% of sales
Consumer profile (Nielsen 2019)	In LR "big buyers" of halal products (4.6% of French households) who make 76% of halal purchases. Young and family profiles: 77% under 50 years old, 46% have children under 11 years old
Share of charcuterie/frozen food (Nielsen 2019)	90% of sales in LRs = delicatessen and salted frozen food Predominance of poultry delicatessen (38%), and frozen meat (17%) 42% of spending on halal products is for charcuterie
Distribution of meat and charcuterie (Nielsen 2019)	80% traditional channels, 20% from LRs

Table 5.6. *Characterization of the halal market and halal meat (Nielsen 2019 and specialized press)*

Main brands/specialized companies	Isla Délice, Isla Mondial, Reghalal, Medina Halal, Halaland, Dounia, Medina, Saada
Main "generalist" companies	Herta, Pierre Martinet, Zakia (Panzani), Fleury Michon, Labeyrie
Private labels	Wassila (Casino), Carrefour Halal
Franchised restaurants	Quick since 2009, halal range present in 10% of the chain's fleet
Perspectives	Development of specialized stores: Halshop and first developments of Halal Bio brands

Table 5.7. *Main halal brands in France (source: data collection from specialized magazines and website/authors)*

Characteristics/ sources	Data
Halal consumption of Muslim households (Ifop 2020)	Systematic for 67% (79% for practitioners and 30% for non-practitioners)
Food budget of Muslim households (Solis 2018)	403 euros/month 34% increase during the month of Ramadan 54% in traditional channels (halal butcher shops, oriental supermarkets, specialized stores) vs. 46% halal departments in LRs
Halal sensitivity by product (% of Muslim households saying they only buy halal) (Solis, Horizons Shopers 2018)	86% meat, 43% stock cubes, 29% bolognese sauce, 24% ready-made meals, 22% soups, 8% baby food

Table 5.8. *Consumption patterns of Muslim households in France*

Features	2014	2018
Number of known halal brands per consumer	6.8	8.7
Share of consumers making purchases based on the guarantee that the product is halal	36%	44%
Share of consumers who are aware of at least three certification bodies	61%	72%
Share of consumers with a preference for the implementation of a specialized halal offer in LRs	59%	71%
Consumer satisfaction with LR halal offer:		
Point of sale	64%	75%
The offer on the shelf	39%	49%
The choice of brands	38%	46%

Table 5.9. *Consumer perception, satisfaction with halal offer and brand awareness according to the 2014 and 2018 Solis surveys*

Characteristics/sources	Data
% volume of slaughter without stunning (Ministry of Agriculture 2016)	15% of cattle and 27% of sheep
Number of slaughterhouses with halal practices (Ministry of Agriculture 2016)	284 slaughterhouses
Certification authorizations	Three mosques: Paris, Evry, Lyon
Certifying bodies	AVS, Achahada, Arrissala, ARGML, ACMIF, Altakwa, HMC, MCI, EHS, EHT, AFCAI, SFCVH, etc.
Performance of the act	700 to 800 certified sacrificers

Table 5.10. *Slaughter and certification of halal meat in France*

5.5.2. *Appendix 2: Description of the questionnaire survey*

The survey was conducted, between January and February 2020, face-to-face using a pre-designed questionnaire. The questionnaire consisted of 32 questions and was divided into five main sections dealing with the purchase and consumption of halal products and halal meat, trust, information about halal products and credibility.

Items	Frequency in % of cases
Age	
18–30 years old	37.78
31–45 years old	32.59
45–60 years old	22.59
Over 60 years old	7.04
Gender	
Male	56.7
Female	43.3
Country of birth	
Maghreb	45.20
France	47.00
Other	7.80
Respondent nationality	
French	71.12
Other	28.88
Parents' nationality of origin	
France	13.34
Maghreb	79.63
Other	7.03

Table 5.11. *Characteristics of the questionnaire survey sample*

5.5.3. *Appendix 3: Example of information collected during the semi-structured interview survey (16 respondents, Montpellier, November 2019–January 2020)*

1. Significance of halal meat consumption (religious act, tradition, ease of purchase, social pressure, fears, etc.)

"The consumption of halal meat is for me a religious act, it is in my traditions. It is the Muslim religion that requires it, it is strictly not a pressure but rather a choice on my part to continue to eat halal."

2. Purchasing and consumption behavior (purchasing channels, frequency of purchase, quantity, consumption occasion, average basket, brands, etc.)

"I live with my parents, so I'm not the one who buys halal meat but rather my parents do. I know that my parents buy our halal white meat from a butcher who always remains the same. For example, the chicken cutlet is fresh and not wrapped. During Eid, my parents get together with friends and buy a veal directly from the farmer. Each person pays an average of 200 euros for 30 kg of meat."

3. Knowledge of the certification process (who certifies? how? existence of a label?, etc.)

"The shopping place is a food store that offers a lot of foreign products (Turkish, Arabic etc.). There is also an area where customers can buy their meat which is certified halal with a label. The certification is done by the three big mosques."

4. What is your assessment of halal products on offer (meat and meat-based products) **in France?** (is it sufficient, qualitative, diversified, etc.?)

"Yes, the supply of halal products is sufficient and qualitative in France."

5. Ease and difficulty in consuming halal meat (availability, credibility, diversity of supply, existence of specialized channels and brands, etc.)

"It is easy to consume halal meat because we already have a regular butcher and even other halal butchers where it is possible to buy our halal meat. We do not have any specific brands."

6. Verification of the halal nature of the purchased meat products (which media? trust? credibility?, etc.)

"The word halal and the certifications are displayed at the butcher's. This proves and reassures us that the meat we buy is indeed halal. The medium used [to communicate this] are paper posters."

7. What does the halal nature mean for the meat? (distribution channel, breeding method, slaughter method, etc.)

"The distribution channel for the veal that my parents buy for the Eid season is short. We know that the animal was raised correctly. The veal goes to the slaughterhouse and is slaughtered in front of the customers while respecting the modalities that Islam imposes, the slaughterer recites the phrase ALLAH AKBAR and then the animal is emptied of its blood."

8. Knowledge and credibility of halal certification bodies?

"We know the halal certification bodies, we can contact them without any problem."

9. Knowledge and credibility of halal product brands and supermarket chains with halal product departments?

"We do not consume halal meat that is sold in supermarkets, even though it is well presented and clean. We are not used to buying pre-packaged meat, and halal meat is usually found next to meat that is not halal and even next to pork. These are the two factors that make us not consume halal meat from supermarkets."

10. Information on halal products (which media? credibility? nature of the information?)

"The instructions come from the three great mosques, which is a guarantee for us. This provides greater confidence."

Box 5.2. *Verbatim responses from the interview of Lina, 18, French, of Turkish origin*

5.6. References

Benabdellah, M. (2017). Understanding the French unconscious for buying halal meat: Local butcher vs supermarket. *La Revue des sciences de gestion*, 3(285–286), 71–75.

Benkheira, M.-H. (2017). Débat : viande de boucherie, rite, fondamentalisme, note de lecture de l'ouvrage de Bergeaud-Blackler F., "Le marché halal ou l'invention d'une tradition. Paris, Le Seuil, 2017, 272." and response from Bergeaud-Blackler F. *Revue Ethnologie Française*, 4(168), 717–728.

Bergeaud-Blackler, F. (2005). De la viande halal à l'halal food. Comment le halal s'est développé en France ? *Revue européenne des migrations internationales*, 21(3), 121.

Bergeaud-Blackler, F. (2017a). *Le marché halal ou l'invention d'une tradition*. Le Seuil, Paris.

Bergeaud-Blackler, F. (2017b). Le marché halal mondial. *Revue du MAUSS*, 1(49), 48–61.

Bergeaud-Blackler, F. and Bernard, B. (2010). *Comprendre le halal*. Liège, Edi.pro.

Bergeaud-Blackler, F. and Bonne, K. (2005). *Premiers résultats d'une enquête sur la consommation halal*. AoFood.

Bergeaud-Blackler, F. and Kokoszka, V. (2017). La standardisation du religieux. La norme halal et l'extension du champ de la normalisation. *Revue du MAUSS*, 49(1), 6291.

Bonne, K., Vermeir, I., Bergeaud-Blackler, F., Verbeke, W. (2007). Determinants of halal meat consumption in France. *British Food Journal*, 109(5), 367–386.

Bonne, K., Vermeir, I., Verbeke, W. (2008). Impact of religion on halal meat consumption decision making in Belgium. *Journal of International Food and Agribusiness Marketing*, 21(1), 5–26.

Boyer, A. (1998). *L'Islam en France*. Presses Universitaires de France, Paris.

Dunoyer, P. (2008). La réglementation applicable dans le domaine de l'abattage rituel. *Bulletin de l'Académie vétérinaire de France*, 161(4), 341–350.

Elseidi, R. (2018). Determinants of halal purchasing intentions: Evidence from UK. *Journal of Islamic Marketing*, 9(1), 167–190.

Falorni O. (2016). Rapport fait au nom de la Commission d'enquête sur les conditions d'abattage des animaux de boucherie dans les abattoirs français. *Assemblée Nationale*, 1(4038), 316.

Godard, B. (2007). *Les musulmans en France : courants, institutions, communautés, un état des lieux*. Robert Laffont, Paris.

IFOP (2020). Les musulmans et la consommation de halal. *L'Express*, Paris [Online]. Available at: https://www.ifop.com/publication/les-musulmans-et-la-consommation-de-halal/.

Lever, J. and Miele, M. (2012). The growth of halal meat markets in Europe: An exploration of the supply side theory of religion. *Journal of Rural Studies*, 28, 528–537.

Meixner, O., Friedl, R., Hartl, B. (2018). Preferences for attributes of halal mat: Empirical evidence from the muslim community in Vienna, Austria. *International Journal of Food System Dynamics*, 9(3), 265–278.

Nielsen (2019). Le marché de la viande halal en France. Report, Buisnesscoot, Paris.

Riera, R., Verliac, F., Berges, J.M., Babre, S., Fenina, M. (2005). Enquête sur le champ du halal. Report, Ministère de l'intérieur et de l'aménagement du territoire, Ministère de l'agriculture et de la pêche, Paris.

Sans, P. and Fontguyon, G.D. (2003). L'industrie de transformation de la viande bovine en France : une approche historique (1950–2003). Report, INRA-LORIA.

Sherwani, M., Afzaal, A., Ali, A., Hussain, S., Gul Zardan, H. (2018). Determinants of muslim consumers' halal meat consumption: Applying and extending the theory of planned behavior. *Journal of Food Products Marketing*, March 22.

Tadiello, C. (2017). La gouvernance du marché de la viande halal. Veterinary PhD Thesis, VetAgro Sup/Université Claude Bernard Lyon I, Lyon.

Wilson, J.A.J. and Liu, J. (2010). Shaping the halal into a brand? *Journal of Islamic Marketing*, 1(2), 107–123.

6

From "Eating Together" to "Living Together Better", the Case of Local Products

René Pierre BEYLIER[1], Fatiha FORT[2] and Andry RAMAROSON[3]
[1] *MRM laboratory, University of Montpellier, France*
[2] *UMR MoISA, Institut Agro-Montpellier SupAgro, France*
[3] *CHROME Laboratory, Centre Universitaire de Mayotte and MRM Laboratory, University of Montpellier, France*

6.1. Introduction

Local products are on the rise! Consuming local products, which is favored by 83% of French people (Merle and Piotrowski 2012), is now a responsible concern for citizens. For the same motivations, 84% of consumers are ready to favor the purchase and consumption of local, home, natural and seasonal products in the future, according to a survey by the Observatoire Cetelem in 2019. Another study published in September 2019 by the GreenFlex firm, Ademe and the YouGov company argues that 80% of respondents have agreed to change their daily practices to reduce the impact of their consumption on their environment. People buying these products are massively convinced that this way of consuming has a positive impact on the local economy (97%)[1], while being in line with the need to return to nature and essential things (81%). As we can see, consuming locally is becoming a fundamental societal trend, especially since cultivation of the land magnifies the

1. Survey conducted by Ipsos and Bienvenue à la ferme in 2014 with a representative sample of 1008 French people aged 18 and above (INSEE quota method: gender, age, occupation of the household's reference person, region and urban area category).

Evolution of Social Ties Around New Food Practices,
coordinated by Gilles SÉRÉ DE LANAUZE.

close and the local, while promoting the sharing of common values and devaluing the distant and the global.

Although this strong interest in local products has continued to grow in recent years, the links between local products and the local have always been close and claimed as such in terms of quality and values linked to the terroir, and also in qualifying the territory, or even eating together. If the links between local products and territory illustrate the evolution of the values and representations attributed to these spaces for more than half a century, it is necessary to note the prevalence of local consumption today, more convivial, more ethical, more responsible, more committed to the territories than global and distant consumption. The multiplication of short circuits, "local" brands, "terroir" brands and "territory" brands, attests to this search for territorial anchoring in terms of proximity, territorial identity, belonging and attachment to the territory, a guarantee of trust, better eating, better health and better living together!

Is this not the objective of the glocalized eater who tries to succeed in two complementary trends that the test of modernization has made contradictory: to be attached to the soil on one hand, and to globalize on the other? In other words, the networked local and the conviviality of the terroir that privileges the links between place, body and spirit "I am what I eat" (Fischler 1990), and the commitment of the digital consumer, who uses global trends in their desire to give meaning to their consumption. The importance of doing something for one's own well-being, contributing to the well-being of others, in other words, making the sharing economy a sustainable economy, favoring the convergence of the values of the collaborative economy and those of sustainable and local development towards a desire to consume differently and better. A locavore and connected consumer, demanding and more enlightened, in search of meaning and trust, authenticity and more transparency, who takes part in the changes and who wishes to change things by infusing new behaviors.

However, consuming locally is not necessarily a guarantee of quality, sustainability or support for the territory. Many actors take advantage of the vagueness to relocalize long channels without allowing the producer–consumer link. Territorial brands are a little too lax and keep the origin of raw materials vague. The difficulty for the consumer is accessing information about the origin and the production–transformation–distribution process. Although this remains opaque for the consumer, initiatives allow us to see more clearly. The "*Ici C.Local*" approach of INRAE is one of them, allowing us to easily identify (color code) local products coming from short circuits and respecting sustainability criteria. Digital applications are being developed to help people live together better by eating locally.

The research problem is then the following: is eating and living better together compatible with eating locally and being a connected consumer? This general question raises two other issues that will be the focus of sections 6.2 and 6.3 of this chapter, respectively:

– Do the local and/or terroir vectors of territorial anchoring (attachment, embedding, rooting), have an impact on the confidence of the consumer-actor and *in fine* on eating together?

– Beyond the motivations to eat locally, what are the factors of the consumption value activated by the perceived image of local products? Are local eaters and connected consumers one and the same, promoting better living together?

To answer these questions, the theoretical framework mobilizes the concepts of: terroir image (Aurier *et al.* 2004), territorial anchoring (Bousquet *et al.* 2020), trust in the brand (Gurviez and Korchia 2002) and in the distributing entity (Kaâbachi 2007), as well as perceived value (Holbrook and Hirschman 1982; Smith and Colgate 2007). The methodology adopted combines interviews and questionnaires (section 6.5, Box 6.2) conducted with consumers of local terroir products.

6.2. Eating locally in a global context

Our eating habits have an impact on the environment – on biodiversity, on climate change – and many advisors and organizations (such as the Hulot Foundation, the Lancet 2019 report) are pushing to "eat local and seasonal". Eating locally seems to be part of a more global trend, more than a fashion, a phenomenon that can be explained by many good reasons for consumption: to preserve the environment of course, and also to preserve one's health, to save money. It is also a way to help local actors and being able to act at one's level on the local economy. More than "made in France", "made in region", "made in territory", "made in terroir" or "made in local" are increasing through locavorism, but not exclusively.

Has consuming terroir products and consuming locally become a societal phenomenon to act globally and together, to reassure oneself and gain confidence? Does consuming locally, which is a guarantee of quality and a vector of territorial anchoring, have an impact on consumer confidence, and *ultimately* on eating together?

After underlining the revival of old local products and observing the shift towards local products and the importance of consuming locally, we will try to analyze how eating local and terroir products favors territorial anchoring and allows us to regain confidence in food products, or even to eat together, based on a study conducted with consumers of local and terroir products.

6.2.1. *Close links between local products and the local area*

Eating locally seems to be part of a sustainable social approach. Consumers are looking for local, natural, fresh, seasonal products, produced or manufactured in an artisanal way – without too much transformation – as close as possible to their homes. Consuming locally is good for the planet, good for the territories and therefore good for the people too. After having defined the terroir and the links with the local from the consumers' point of view, we will show how consuming locally is more committed to the territory.

6.2.1.1. *From terroir to local: diversity of products promoted by these spaces*

The food productions for which the geographical origin is identified are of great diversity. Poulain (2018) suggests approaching all of these productions linked to a place and which have a historical and cultural depth from three criteria, since they cross space and time, and are the object of shared know-how.

6.2.1.1.1. Terroir and local products

The expression "terroir" seems to have always existed[2]. Its origins date back to the French Revolution, but the association between "products" and "terroir" is much more recent (Bérard and Marchenay 2004). Defining the concept of terroir has occupied many researchers in agronomy (the land, the exhibition, etc.), sociology (social construction of territorial quality), geography (physical and human space), anthropology (representations, symbols) and economics (quality signal, reputation, etc.). The INAO[3] proposes a definition based on a very technical and operational

2. The word "terroir" appears in the treatise "*Théatre d'agriculture et mesnage des champs*" (1600, 1024 p.) by Olivier de Serres (1539–1619), faithful agronomist to King Henri IV (1553–1610).

3. INAO: *Institut National de l'Origine et de la Qualité*, created in 1935, is a public administrative institution under the supervision of the Ministry of Agriculture and Food, responsible for implementing French policy on official labels of identification of the quality and origin of agricultural and agrifood products (AOC, AOP, IGP, LR, AB, etc).

approach that serves as a basis for the certification system of appellations of origin and for an international definition[4]:

> A terroir is a delimited geographical area defined by a human community that has built up, over the course of its history, a set of distinctive cultural traits, knowledge and practices, based on a system of interactions between the natural environment and human factors. The know-how brought into play reveals an originality, confers a typicality and allows recognition for the products or services originating from this space and therefore for the people who live there (Casabianca *et al.* 2006).

Beyond this "technical" definition, we note that for consumers, a terroir evokes an ideal of good food, countryside, land, roots, traditions and commitment to supporting local producers. The value of terroir products also depends on the "intangible work" invested by marketing professionals in building a brand image focused on product quality and authenticity (Quellier 2018). This discursive construction of a terroir as a guarantor of quality and authenticity is not new and has existed since the beginning of the Fordist era (Quellier 2018).

Historical studies have clearly shown that the typicality of terroir products results above all from their ability to travel, to be exported outside of their territory of origin and to meet outside consumers (Delfosse 2012). Thus, the association between product and terroir is the result of a long and complex construction process in time and space (Bérard and Marchenay 2004), in other words, the result of the interaction between the product and its place of production/manufacturing. In addition to their geographical origin, local products convey an often long history and traditional and authentic know-how. They are distinguished from other food products by their strong identity and their distinctive characteristics, such as very typical organoleptic characteristics, a very present social history, a standardized and collective artisanal type of production and a high price, guaranteeing superior quality.

The polysemous definition of the concept of a terroir is not likely to simplify the expression "*produit de terroir*", especially from the consumer's point of view. Although in their minds terroir products are often associated with local markets, they are not reduced to this. Consumers consider a terroir product to be local when its origin is local or when there is geographic proximity between the place of production and consumption of the product (Feldmann and Hamn 2015).

4. INAO-UNESCO – "Rencontres internationales Planète Terroir", Terroirs et Cultures, 2005.

6.2.1.1.2. Local products and localized products

Local, like terroir, is a polymorphous concept. The definition of a local product could be justified by the link between a food product and the space in which it is produced/manufactured and/or consumed, as attested by the media, activist and scientific interest, particularly in short circuits (direct sales at the farm, collective sales points, etc.). Although there is no regulatory definition, a consensus is emerging around the concept of geographical proximity, the distance between the places of production and consumption of the product. The literature distinguishes localized products from local products (Merle and Piotrowski 2012).

Localized products refer to a geographical origin (name of the place of production such as "Ventoux cherries", or the AOC, AOP, IGP labels), identifying an agrifood product that draws its authenticity and typicality from its territorial origin. The product is said to be "localized" when it is consumed in a place other than its production area. According to this interpretation, a local product whose origin is guaranteed – originating from a terroir – and which is consumed in a region different from its place of conception can be considered as a localized product. This vagueness in the mind of the consumer is widely maintained.

Local products are differentiated from localized products by the geographical proximity – a space that is measured in terms of physical distance, time and access, and even transport costs – that exists between the production-processing activity and the consumption activity (Merle and Piotrowski 2012). Originating in this identified space, they are attached to a local territorial origin, and they are defined as products distributed and consumed in this same space or close to their place of production-manufacturing, unlike localized products. A maximum geographical distance of 100 kilometers between these two places has been indicated by Feldmann and Hamm (2015). For other researchers, local products refer to a situated space whose limit seems to be fixed by concentric operation according to the type of product sought, the level of processing (supplying a school canteen with poultry locally is possible, but not chicken breast in large quantities). In other words, the so-called "local" space is at the discretion of the consumer and the actors in the territorial field.

Thus, "terroir" is not equivalent to "local" or "localized". However, there are obvious links between the concepts of terroir and "local", and between terroir and "localized", notably through the development of a territory and support for it, and the construction of its identity through the consumption of food products.

6.2.1.2. *The prevalence of local consumption, more committed to the territories*

The link between a product and the territorial space in which it is produced-manufactured and/or consumed is the subject of debates around local products, local and/or localized products, short circuits, direct sales at the farm, in companies or other collective sales points (Malié 2016)[5]. If confusion reigns between the terms qualifying the products "from here", for all that, the marketing draws advantages and benefits from it. Consuming locally is an increasingly present citizen concern, and refers to multiple motivations for buying local.

6.2.1.2.1. The terroir "confined" to the local

Although "terroir" is not equivalent to "local", in the literature and in surveys, "terroir" and "local" products are used indiscriminately. However, if certain local products can be considered as local (Roquefort cheese is local for the inhabitants of the Cévennes, but is not local for the inhabitants of Lille), this is not the case for local products which do not systematically refer to the terroir. Today, we are witnessing a shift from terroir to local (Box 6.1).

The locavore movement, with initiatives to connect consumers to their territory, is largely supported by the communication campaigns of local authorities around local territorial brands. However, we note that the local level is of variable geometry, that of urban communities and agglomerations, that of project territories designed by the regions or by Europe (Pays, Local Action Groups of Leader programs, etc.), or those designed by the new metropolises (Delfosse 2012). Project and management territories rarely correspond to a terroir in the sense of appellations of origin.

Today, there is a tension between terroir-nature and local, between local and global. Indeed, the concept of terroir products has been globalized since the WTO TRIPS agreements in 1994 and is still very popular, especially in developing countries, encouraged by international development organizations (FAO, UNIDO, ADB, etc.), which see it as a way to improve producers' incomes and product quality. Global markets for wine and other local cheeses are developing in emerging countries such as China and Japan, but the trend towards localism is present in consumer purchasing behavior everywhere. It is therefore possible that, after having left the localized peasant economy and conquered markets that are sometimes very far away, local products will experience a shrinking of their marketing territories.

---

5. Malié, A. (2016). C'est local, c'est ce qui nous intéresse : étude des constructions et usages du "local" à travers les pratiques alimentaires. In *Les territoires de l'autochtonie*, Aunis, E., Benet, J., Mège, A., Prat, I. (eds). Presses Universitaires de Rennes, 97–110.

Birth of the terroir concept

Although Parker traces the gestation of the concept of terroir back to the 17th century in the history of French gastronomy, institutionalization dates back to the 1919 law that defined wines by their origin and allowed the constitution of defense unions (Quellier 2018). The local products became the emblem of the country thanks to intensified tourism, leading to the patrimonialization of the land.

Between the wars and after the war

Delfosse (2012) considers that the interwar period saw claims for product diversity "conceived as a mirror of national cohesion and as a means of ensuring a lost French supremacy on the international political scene". The post-war period and the efforts to rebuild the country and modernize agriculture eclipsed local products from markets and restaurant tables. The development of mass distribution and industrialization tended to standardize products and reduce the diversity of food on offer.

From the 1980s onwards

From the 1980s onwards, in connection with the questioning of the agro-industrial model and the imperatives of local development and protection of the rural heritage, there was a proliferation of requests for designations of origin to protect local products from counterfeiting and the relocation of production. We can see the strong infatuation of producers, industrialists and distributors with the concept of local products, reinforced by food crises that push consumers to seek the origin of products as a means of reassurance.

The 2000s

The advent of sustainable development and protection of the planet put the terroir back in the spotlight. Local products appeared to be sustainable, close to nature, respectful of the seasons, etc. However, consumers concerned about the consequences of their food on their health and that of the planet were beginning to look for "the close, the local, which no longer necessarily values the unique, the exceptional, but also the everyday, the banal, the product that does not necessarily have any other geographical attribute than that of local" (Delfosse 2012). With the locavore movement, for French consumers, terroir was becoming everyday and concerned products other than wine and cheese, those with no name, nor claiming any singularity, except that of being from here. This trend was reinforced by decentralization and the responsibility of local authorities for the economic development of their territory. Through communication campaigns such as "buy local" or "your purchases make our jobs", territorial brands have flourished all over France, supported by cities or regions.

Box 6.1. *From terroir to local*

6.2.1.2.2. Local products, a large family for the consumer and a marketing asset

Local products, local food, products of the terroir, typical food, regional food, food of local origin or even regional heritage, this clever mix of all of these terms reminds us of the concept of territory, of locality, of terroir linked to food culture. They would then be defined as a black box that Latour[6] conceives as a stable network that becomes obvious and that nobody questions. These products form a vast family that includes all sectors of the agrifood industry: unprocessed plant and animal products and processed products such as ready-made meals, charcuterie and confectionery. They are expressed through the link between food and culture. They are expressed by the more or less close link with their local society, and the richness of the production, distribution and consumption methods. The local product, like the terroir product, is not a static concept; it is multidimensional, evolving and adaptable to the types of consumers. This conceptual limit is a marketing asset for the organization that will be able to take into account the expectations of consumers and their own interpretation of the word "local-terroir", depending on their involvement.

Marketing plays on these differences and this confusion in the consumer's mind between terroir, local, localized, and even origin and provenance, to promote local products produced hundreds of kilometers away (Casino's locavorism: *Le meilleur d'ici*, the group's own local brand). If the local consumer does not necessarily need to identify the product as "from here", the explicit link to the territory has an undeniable commercial role and a competitive added value outside the region of origin.

6.2.1.2.3. "Consuming local", motivations and commitments in favor of the territory and better living together

According to Paddeu (2017, p. 40), consuming locally aims to "reduce the impact of the agricultural and food system on biotic and abiotic communities[7], an alternative practice designed to short-circuit globalized capitalist modes of production, the invention of a space of experimentation where other ways of cultivating, exchanging, and feeding ourselves take shape". Eating would then be a way to reweave the links between all of the stakeholders in the territory, particularly between producers, distributors and citizen-consumers. Is the local eater more active and engaged, more responsible towards the territory and its actors? What are their motivations?

6. Latour (1987). *Science in Action*. Harvard University Press, Boston.

7. Interaction between a group of living organisms (biotic communities) and inanimate aspects (abiotic environment), brought together in a natural way, living together in the same space.

The literature review allows us to extract two types of motivations that drive consumers in the choice of a local product: individual motivations of a functional, health and/or economic nature and collective motivations that are more committed to the territory and better living together, such as sharing, commitment and social value (Table 6.1). Many authors on consuming local (Merle and Piotrowski 2012; Malié 2016; Arsil *et al.* 2018) highlight consumers' expectations on product quality, less intensive and non-industrial modes of production and manufacturing, respectively. Similar to previous research, Arsil *et al.* (2018) mention that local foods are considered to be of higher quality than non-local foods. They are rated as fresher, more nutritious and superior in taste, smell, and appearance in terms of color, shape and texture, than non-local products. Among other preferences and attributes highlighted by Arsil *et al.*'s (2018) work, food safety has a greater influence on urban consumers of local products. Consumers view local products as being produced and/or manufactured using traditional skills. These values of local tradition stem from the belief that what has been tried and succeeded in the past, what has stood the test of time, is still valid, good, and therefore reassuring.

The literature (Table 6.1) clearly shows that eating local is a desire to consume in a responsible and shared manner by considering two axes – the social and the environmental (Merle and Piotrowski, 2012). On the social level, while contributing to keeping traditions alive and ensuring that they are passed on (Bowen and Mutersbaugh 2014), eating local allows the local and/or territorial economy to be sustained by supporting producers, craftsmen and SMEs near one's home, by recreating social links between local actors. These concerns and motivations for solidarity are the result of a society in discomfort or crisis: fear of unemployment, an upsurge in bankruptcies and suicides among farmers, difficulties and bankruptcy filings by SMEs, and so on, as well as the rejection of globalization and relocation (Merle and Piotrowski 2012), a form of antidote to standardization according to Bowen and Musterbaugh (2014). In terms of the environment, this type of consumption is seen as: the choice of an activity among suppliers (producers, artisans, SMEs, etc.) with a non-intensive or low-industrialized manufacturing process, support for alternative production/distribution models close to the place of consumption in order to reduce the carbon footprint, preservation of biodiversity and landscapes, safeguarding of culinary heritage and traditions. Eating local is therefore a committed and responsible choice, and it is to consume ethically; in other words, it is to have a consumption that is both respectful of the environment and of cultural diversity, beneficial for the local economy, good for health and positive for society. This form of consumption can be defined as a behavior carried out by consumers who are concerned about the environmental, socio-economic and societal problems of the territory. Concerned about the negative consequences of these problems on themselves (selfish motivations) and on others (altruistic motivations), they opt for local products, festive products, labeled products whose origin is guaranteed, less toxic and more sustainable products. Altruistic values reflect a concern for social justice and societal well-being, a way of sharing to live better together. Distributors are not mistaken and in order to respond to the motivations of consumers, such as the Biocoop[8] brand, are engaging in actions that highlight the territories or regions of France by emphasizing the importance of the relocation of channels. Local products are considered to be top-of-the-range products, and festive products that are shared with others, a moment of conviviality around "eating together".

Motivations of order		Values and commitments associated with local consumption in favor of better living together
Individual	Functional	Improving consumer perceptions of organoleptic attributes (taste, freshness, superior quality, naturalness and healthiness)
	Health	Adopting a healthier lifestyle by reducing health risks and negative health consequences of environmental problems
	Economic	The right to access quality, healthier food with personal benefits (health, symbolism and status) at a fairer price
Collective	Altruistic ethics	Supporting local producers and the local economy, the regional economy (maintaining jobs, better remuneration, etc.), solving environmental and societal problems: concern for socio-economic justice and the well-being of others
	Social	Search for a social link with local actors; territorial solidarity; adhesion to a distribution system; search for relational proximity; more active, responsible and shared participation
	Socially responsible consumption	Consuming responsibly, preserving nature and solving socio-environmental problems; myth of the natural

Table 6.1. *Motivations to consume local products in favor of the territories and better living together*

In the rest of the chapter, we will talk about local-localized and terroir simultaneously or undifferentiated like the consumer.

8. Biocoop proposes its project for the future with *Hungry and Foolish*, a TV film broadcast on the air and on YouTube.

6.2.2. *Food and local anchoring: the challenge of trust*

The paradox is significant. Although French food remains one of the best in the world in terms of organoleptic and sanitary qualities, and controls each stage of production–manufacturing–distribution, confidence in it seems to be eroding. Consumers are increasingly interested in local products, in sharing their meals with each other and on social networks. Could eating locally be the vector of a newfound confidence and anchoring? After underlining the consum'actor's distrust of food, we will try to measure the impact of the terroir image on confidence in this type of product, based on a quantitative study conducted among consumers of terroir products.

6.2.2.1. *Eating locally: origin and transparency to overcome mistrust*

Food is undergoing a profound revolution. Recent food scandals (lasagne with horse meat in 2013, organic eggs contaminated with fipronil[9] in 2017) and health crises have contributed to the creation of a strong feeling of mistrust towards the agro-industry. If they are powerful revelations of our food dependencies, they are also strong stimulators of the mutations in progress, particularly food transition[10]. One in two French people often feel as though they no longer know what they are eating (Ipsos and Bienvenue à la ferme[11], 2014). Moreover, more than 91% have declared that they are concerned about what they eat (Study of the Consumer & Retail pole of Ifop, 2018 "*la consommation alimentaire des français : évolutions & attentes*"). Out of mistrust, two-thirds of respondents have abandoned their usual brand, believing that these companies do not respect them (70% according to the Kantar TNS group, January 2018), and are increasingly favoring local products. Local identifiers are important to them, and brands, signs and labels that certify both the origin of products and/or ingredients and also transparency on production–manufacturing–and/or distribution methods are more sought after. Consumers remain wary of large food brands and have more confidence in local producers and original products (Table 6.2).

---

9. Fipronil, an insecticide not approved for use in food-producing animals, is used for the control of lice, mites and other parasitic invertebrates found in domestic animals, and also used in the food chain as a crop protection agent.

10. All of the future changes in consumption habits and, in parallel, in production methods due to the evolution of consum'actors' expectations.

11. Bienvenue à la ferme is a network of farmers federated by the Chambers of Agriculture. Created more than 25 years ago (in 1988), Bienvenue à la ferme, later joined by the Marchés des Producteurs de Pays, has become the first national network of direct sales and farm reception.

Trust/confidence by product, label and brand	% no confidence
Food products from small producers	18
Quality labels (Label Rouge)	33
Food products guaranteed to be of French origin	33
Products whose ingredients are guaranteed to be of natural origin	43
Organic food products	49
The major food brands	63

Table 6.2. *French consumers' distrust of food products (source: Opinion Way for French Food Capital Oct – 2018 (LSA November 18, 2018))*

Crises often trigger or reveal new strategies and behaviors. The mad cow crisis (end of the 1990s) led to an increased distrust of state guarantee systems, which appeared to be fallible. Consumers then sought to know the origin of products or went directly to the producer. Today, eating locally is very present (more than one in four French people are adept at eating locally) and various forms of consumption have emerged: an interest in open-air markets, farmers' markets, non-committal short circuits (frequented by young people, the less affluent classes) unlike the more committed and militant AMAPs, cooperative supermarkets and via Internet platforms.

Studies conducted with French consumers show that among the attributes mentioned above, origin and transparency seem to be of primary importance to consumers. They place a high value on origin, 53% say they pay more attention to provenance than they did before, and 97% say they read the list of ingredients on products. Transparency is another consumer concern (Table 6.3). A demanding eater, in search of meaning, authenticity and more transparency in practices (according to Feldmann and Hamm (2015)), is an eater attentive to production–manufacturing conditions (animal welfare, use of pesticides, etc.) and exchanges in the supply and distribution chain (Bowen and Mutersbaugh 2014).

Local is often associated with short circuit. The short circuit does not include the notion of distance; it is only limited to a maximum intermediary between the producer/manufacturer and the consumer. However, these are indeed local circuits that are being developed under the impetus of local authorities. Paddeu (2017) studied the re-localization of food systems in order to transform and reinvent them. A virtuous local, certainly, but a subversive local that can become a lure if it is based on "the interests of a restricted, even authoritarian elite" (Paddeu 2017, p. 43).

Importance of the criteria granted by French consumers			
According to Nielsen (2019)	**%**	**According to Kantar TSN (2020)**	**%**
Local product	82	Respect for animal welfare	91
Product made in France	82	Respect for the environment	90
Regional product	78	Less packaging	86
Handcrafted product	76	Poorly paid farmers	73
Local product	74	Desire to eat healthy	63
Product with a label	70	Desire to eat a balanced diet	59

Table 6.3. *Importance of origin and transparency in French consumers eating locally*

Consumers are now, more than ever, very attentive and rather suspicious about their food, even though it comes from local products. Among their main preoccupations as local eaters, transparency is one of the major concerns, along with origin, in order to regain trust. This is the subject of the next item based on the supposed link between the terroir image of products, their perceived value and trust, from a field study conducted with consumers.

6.2.2.2. *Local and regional products, a guarantee of quality and renewed confidence*

Local food is seen as a way to address "trust deficits", in the words of Holt (2007). The terroir, the local, favored by the consumer, would contribute to strengthening the concept of "food sustainability" (Holt 2007). This concept concerns both the product and the processes of production, distribution and consumption, processes that use artisanal methods of ecological production, short supply chains and home-made and seasonal cooking. Localized products would promote territorial anchoring based on the joint proximities mobilized. These allow for the creation of relationships of trust between stakeholders. *The establishment and maintenance of trusting relationships also requires that organizations in the production–distribution chain be perceived as transparent in their operations.* Trust is explained in particular by the reputation and legitimacy that emerge from territorial anchoring, according to Bousquet *et al.* (2020), through the consumption of local terroir products. The terroir, the local and the products they carry confer an image that is favorable to the reinforcement of trust between actors and what they eat.

The literature shows that the product brand image construct helps create a sense of belonging (Charton-Vachet and Lombard 2015) and promotes trust, while creating value for the company. Lenglet and Müller (2016) found a positive significant relationship between elements of localized product image (geographic proximity), consumption experiences and consumer trust. Thus, we suggest the following hypothesis: the perceived image and the perceived value of localized and/or terroir products have an impact on the consumer's trust in the brand and in the organization that produces and/or distributes them.

The quantitative field study was conducted on a sample of 1034 French consumers of local products (section 6.5, Box 6.3, Tables 6.11 and 6.14). The exploratory factorial analysis allows us to identify two components (Table 6.4): origin, profession and culture (OPC), on the one hand, and time and naturalness (TN), on the other hand. These results are quite similar to those of Aurier *et al.* Do the characteristics of the terroir image perceived by consumers have an influence on trust?

Built	Dimensions	No. of items	Jöreskog's rho reliability	Rho convergent validity
Terroir image of the product	Origin, profession and culture	4	0.885	0.659
	Time and naturalness	4	0.884	0.657
Factor contribution between the two dimensions: 0.808 – t-test = 9.981 (>1.96) with $p<0.001$				

Table 6.4. *Results of the factorial analysis of the perceived terroir image of products*

The literature shows that trust is either a logic of calculation and interest (individual interests in maintaining one's reputation and/or legitimacy, and also a risk, the transaction refers to a cost/benefit calculation), or the expression of a social and collective reality, in other words, a belief of individuals in a relationship that depends on the networks developed and the embedding of the socio-economic spheres (Granovetter 1985). In this last perspective, the most relevant trust results from contextualized and socially embedded social constructions (Granovetter 1985) within the territorial field. The characteristics and foundations of trust evolve according to the context of the relationship, and are therefore an integral part of the history of this relationship between the partners. The relationship of trust with producers–manufacturers of local products is one of the characteristics regularly put forward by consumers, who are looking for products that meet both functional (utilitarian), experiential (hedonic) and symbolic (expression and responsibility) needs. Eating local and terroir food gives meaning to consumption, allows consumers to establish a social link and to reconnect with local actors, and through "eating together" highlights the notions of conviviality, sharing and social value expressed by the experiential and symbolic dimensions of the perceived value of consumption (Table 6.9). Thus, terroir and "eating together" through the characteristics of "terroir origin" and perceived consumption value would influence consumer confidence in the brand and in the entity that distributes it according to the channels used.

Brand trust, defined by Gurviez as "the presumption that the brand, as a personified entity, is committed to predictable action consistent with its expectations, and to benevolently maintaining this orientation over time", is measured by the Gurviez and Korchia (2002) scale with three dimensions: credibility, integrity and benevolence (section 6.5, Table 6.12). Trust in the distributing entity is measured by the scale developed by Kaâbachi (2007) based on three items corresponding to the three dimensions of the Gurviez and Korchia scale.

6.2.2.2.1. Results of the analysis: Terroir image and perceived value as antecedents of trust

The factor analyses – exploratory and confirmatory – and the tests carried out confirm the validity (factor weight greater than 0.5) and internal consistency of the scales used (Jöreskog's rho greater than 0.7) (Table 6.5).

Built	Dimensions	No. of items	Jöreskog's rho reliability	Rho convergent validity
Perceived value of consumption	Experiential	7	0.948	0.724
	Symbolism	6	0.937	0.713
	Functional	5	0.897	0.635
	Ethics	4	0.884	0.658
Brand confidence	Credibility and integrity	5	0.912	0.677
	Kindness	3	0.874	0.698
Trust in the sign	One-dimensional	3	0.896	0.742

Table 6.5. *Reliability and validity of the perceived value and confidence scales*

The tested relationships "terroir/local image on trust" are all positive and valid (Table 6.6). The "terroir" characteristics, linked to the local product, have a positive effect on trust in the "brand" identified as being of territorial origin. The two components of the terroir image – Origin–Profession–Culture, and

Time-Naturalness – have a significant positive influence on confidence in the product. The same is true for the "ethics" variable, which helps to develop trust.

Relationships tested (Multiple RL)	Std Coeff. β	Test t	Sig. p
Terroir/local image → Brand confidence	0.678	17.261	0.000
Origin-Profession-Culture → Brand confidence	0.427	7.293	0.000
Time-Naturalness → Brand confidence	0.340	5.359	0.000
Ethics → Brand confidence	0.311	5.387	0.000
Confidence/brand ↔ Confidence/distributing entity	0.670	16.866	0.000

Table 6.6. *Validation of the effect of terroir/local image on trust*

Relationships with local food, proximities built on cooperation, common interests and ethical and civic commitment are a factor of trust following the example of the work by Bowen and Musterbaugh (2014). Consuming locally sheds light on the importance of the place of origin and is based on a logic of proximity between the producer and consumer (Dufeu and Ferrandi 2011) through the perceived image of origin, a proximity that goes well beyond mere location, exploring, in particular relations with stakeholders, the craft, in other words, the processes implemented in the manufacture of the product (inherited traditional know-how), identity, that is, adherence to the values represented by the product attached to a place (history, tradition, naturalness, etc.) and the entity that designs and/or distributes it (ethical value).

Perceived value is also a determinant of brand confidence, as the positive and valid links in our study attest to (Table 6.7). The notions of conviviality, sharing and social value through the experiential and symbolic dimensions of consumer value promote brand trust, justifying the link between terroir and trust. Confidence in the brand can then be considered as a marker of eating together, or even of living better together with the ethical component (Table 6.1).

Relationships tested (RL Multiple)	Std Coeff. β	Test t	Sig. p
Perceived value → Brand confidence	0.712	18.997	0.000
Experiential → Brand confidence	0.447	6.549	0.000
Symbolic → Brand confidence	0.161	2.910	0.002

Table 6.7. *Validation of the perceived value effect on brand confidence*

Furthermore, confidence in the local product has a significant positive influence on confidence in the producer and/or distributor and vice versa (Table 6.6). The managerial interest of this study is threefold. It shows the interest and complexity of the influence of the local product image on the behavior of the local eater. Second, it demonstrates the role of the territorial network (plural actors) in the promotion of confidence in the product, which requires a better understanding of the forms of association between actors and implies transparency in the production–distribution chain. However, commitment in the networks (associative or not) favors better living-together[12]. Finally, it underlines the diversity of the motivations of the local eater as well as the need to pay particular attention to the role played by territorial anchoring and proximity inherent to this type of product and/or circuit and by trust on the commitment of the local eater.

6.3. Eating locally: from local conviviality to globalized connectivity

Eating is "a total human act", as Morin (1973) asserts[13], a social and cultural fact, as well as a geographical fact, as Fumey (2010) demonstrates in an essay on cultural geography. The meal is "at the heart of socialization, a place to learn the rules of life in a group and a place of sociality, sharing, conviviality" (Poulain 2018). The eater is also a connected citizen-consumer, which gives them access to multiple points of contact at any time and in any place. The literature notes a possible "dehumanization of relationships" (Rogeon *et al.* 2019) as a consequence of these virtual relationships. Yet, shopping for local products lies in the proximity to territorial actors, from the producer to the distributor and other customers. Rogeon *et al.* (2019) provided a grid for reading the paradoxes between digitalization and local product shopping.

In section 6.2, we showed that the attributes of the terroir image of localized products are sources of trust for the local eater, confirming the thesis of Merle and Piotrowski (2012), and underlined among the multiple motivations for eating local, the importance of origin in particular, and also of: proximity, transparency, the desire to consume responsibly and the search for a social link. The objective of this second part is to identify the values sought in the consumption of local products, and then to measure the effect of the dimensions of the image of these products on the

12. This is the meaning of the "Colibri" movement founded by Pierre Rabhi, like the hummingbird that tries to extinguish the fire that is ravaging the forest by carrying a small drop of water in its beak, the consum'actor tries to "do their part" in their exchanges with other actors in the territorial field.

13. *Le paradigme perdu : la nature humaine*. Seuil, Paris.

perceived value, from a field study. Finally, we will attempt to verify the compatibility of the local eater and connected consumer through the literature.

Of the two sections developed below, the first shows that the terroir, the local, although rooted in tradition and sharing, are also a place of innovation and projects. It highlights the polysensualism of the local eater through territorial anchoring, promoted by the terroir image of local products highlighted by a mixed qualitative and quantitative study. The fieldwork attempts to answer the question: what are the facets of the perceived value of consumption activated by the attributes of the terroir image of local products? The second section, through a review of the literature, shows that digital technology is revolutionizing the daily lives of individuals, particularly in terms of sharing goods, services, values and commitment, and is impacting their lifestyles, including their way of sharing moments of conviviality. The habit of carrying out several tasks simultaneously makes it possible to eat together by standing or sitting in several successive places, sharing a screen. Digital technology is an opportunity to live together better and to relocalize: to meet, to gather, to exchange and develop a project together, to identify the local offer to consume locally and to contribute to the development of one's territory. A review of the literature provides answers to the following questions: can the local eater become attached to a territory while being connected to the world? If digital technology allows for the creation of links between people who are geographically close, does it serve the purpose of better living together? In other words, are digital technology and better living together through local consumption compatible?

6.3.1. *Attachment to a soil: the conviviality of the land and better living together*

Eating behaviors are changing, the link between food and health and wellness is well established, and expectations differ according to generational cohorts and context. But each generation keeps a certain number of habits, brakes and attractions that are specific to it in mind. Motivations are multiple and remain relatively personal, the consumer of this type of product wishes to enjoy pleasure and share this pleasure with others. This rise in hedonism and emotions is reflected in the search and/or rediscovery of gustatory quality in particular, but not exclusively. The polysensualism of the local eater emphasizes the importance of senses other than taste and the pleasures of a convivial table: sharing emotions around eating and drinking, valuing local cuisine (authenticity), concern for health (for oneself, one's children and others), making new experiences of sharing and solidarity, or the pleasure of rediscovering one's roots (nostalgia), and showing one's attachment and/or belonging to the territory. The construction of the link to the territory through local consumption seems to convey various forms of territorial anchoring between representation and standards (certification) in regard to the links to the market and to the product: construction of a quality linked to the associated traditional know-how and also to innovations, and a social form resituating the product in its food cultures.

The literature review and the mixed study – interviews and surveys of local terroir products consumers – show the importance of consuming locally to eat together and live better together through territorial anchoring and perceived value.

6.3.1.1. *The local network: between tradition and shared innovation*

If globalization is synonymous with standardization, even impoverishment from the point of view of the general public, it is comforting to read the analyses of researchers on the role of innovation in local socio-economic development. A trend that relies on the notion of local traditional resources, historically constituted – the terroir – but in continual construction, which reinvents itself, readapts and reformulates itself with each transformation of the context. Indeed, the creation of a localized system of production, transformation and distribution is part of an innovation process. We do not return to past methods, nor do we go backwards. Therefore, the production of terroir, localized products, AOC and AOP standards protecting a know-how anchored in a history and a tradition are recent collective constructions that ensure a superior and specific quality to the representations of producers and consumers alike. The appellations of origin are an example of innovation of tradition, of dynamic (innovative) construction from ancient know-how, uniting all of the actors – from the producer to the consumer – of a defined territory in a solidarity network. The notion of a localized system implies and requires the ability to appropriate and develop new processes – products, markets, distribution, consumption – based on the link to the territory.

6.3.1.1.1. When the local network innovates: co-construction of actors

Is the local product a matter of collective ownership and innovation? For the anthropologist Hubert (2001)[14], the territorial product is a product of interaction between producers and consumers, where products and individuals are identified, as well as a form of innovation in the production, quality and distribution processes. The product is territorially anchored, since the resources, which are the result of the work of local actors over time, are the result of a succession of collective actions and the pooling of skills and evolving know-how. The joint geographical and organizational proximities allow for the creation of relationships of trust between the different partners (Bousquet *et al.* 2020). The literature shows the importance of

14. Annie Hubert (1941–2010), a great anthropologist of food and health, was director of research at the CNRS.

actors in the construction, functioning and development of territorial markets, and in particular of intermediaries (such as coordinating agents, agents of chambers of agriculture and consumer associations). Their effectiveness is based on the trust established between the actors involved in their coordination, information and training actions (e.g. training farmers to the market), in other words, allowing the confrontation of the aspirations of producers with the economic realities, the reality of the market, in response to the expectations of the consum'actors. This networking of actors, at the origin of certain short and proximity circuits such as the market of Grabels in the Hérault or that of Carpentras in the Vaucluse (markets reserved for producers in short circuits), reconstitutes the social links forged between producers, consumers, intermediaries, organizations and institutions at the same time. This results in tough negotiations, compromises, innovation in the processes, relocation of supplies (transparency, origin) and an innovative reappropriation of know-how, allowing the processes to become anchored in the practices of the different actors.

6.3.1.1.2. Territorial anchoring, networks and innovation

Territorial anchoring is the expression of a territorial identity, of a trend that stands in opposition to globalization, thanks to the mobilization of networks of action and collective efficiency, such as family networks, professional networks and institutional support networks, all of which are closely interwoven. The literature on territorial innovation systems shows that social networks, which bring together heterogeneous actors who are simultaneously open on different scales, make it possible to stimulate innovation through their capacity to mobilize around a common project. The neo-institutional sociology approach[15] encourages organizations to broaden their field of vision to include all of the actors in the organizational field[16] in which they confront, exchange and socially and collectively construct the "social meaning of reality". The social meaning of reality is based on a set of shared meanings, such as cultural heritage, history, identity, territorial belonging (Charton-Vachet and Lombard 2015), symbols, regional feeling (Dion *et al.* 2010) and food cultures. This approach highlights the role of networks and the importance of proximities within the territorial field in the ongoing construction of food cultures. The local territory in question, and *a fortiori* the terroir, are spaces marked by the idea of proximity, collaboration and cooperation (Pesqueux 2015), a

15. The NIS (DiMaggio and Powell 1983) posits that environmental forces drive all actors in the same arena (the same organizational field) to adopt and/or share the same collectively constructed social and cultural reality.

16. The organizational field according to Scott (1995): "*a community of organizations that has a common system of understanding and whose participants interact more frequently* [...] *with each other than with actors outside the field*".

place of resources and common construction. Although the link to the local is a link to a natural and cultural heritage, to the imaginary, to social and symbolic representations, tradition is readjusted in the common project of valorization. Dedeire and Tozanli (2007) speak of a process of mutual and paradoxical acculturation in the encounter of man with other cultures. The memory of a taste, a smell, a know-how, an experience moves with man, accompanies them. Thus, humanity spreads original culture to others and is also impregnated with cultures near and far. The local products, whose origin is highlighted, are carriers of symbol, of tradition[17]. For Pesqueux (2015), tradition refers to the weight of the living past (there is memory in tradition, he says). Transmission, inheritance of cultures and tradition, certainly, but with the idea of integration of new elements, hence Pesqueux's (2015, p. 106) formula, "*There can be no innovation without tradition*". Like tradition, innovation is built on principles (debatable and not very evolving), and norms (evolving). Innovation is then put forward "*as likely to solve problems related to sustainable development, health crises and social difficulties*" (Pesqueux 2015, p. 100).

The consumption of local products, whose origin is emphasized, is used to convey cultural meanings (Arnould and Thompson 2005). For McCracken[18], the culturally constituted world, consumer goods and the individual consumer represent the three sites of meaning. Thus, the culturally constituted world, representing the origin of cultural meanings, transfers meanings to the product-objects through rituals and customs. This transfer of cultural meanings to individuals allows for a better understanding of how traditions and lifestyles are transmitted (Arnould and Thompson 2005) and enriched by the integration of innovations promoted by new individuals inserted into the local group and/or through networks. The territory favors this insertion, aiming to reduce distances within and increase them with the outside (localization vs. globalization).

6.3.1.1.3. The consumers' discourse

Semi-structured interviews conducted with about 30 consumers of local and regional products in southeastern France were undertaken to understand and characterize the act of purchasing and consuming this type of product related to their territory (section 6.5, Table 6.11).

17. Tradition, in Latin *traditio*: "*act of transmitting*", comes from the verb *tradere*, "*to pass on to another, to hand over*" (Alleau, Encyclopedia Universalis).

18. McCracken G. (1986). Culture and consumption: A theoretical account of the structure and movement of the cultural meaning of consumer goods. *Journal of Consumer Research*, 71–84.

	Characteristics related to territorial anchoring / consumer discourse*
Importance of socio-economic links (embedding and living together)	Links forged and established between producers and distributors
	Exchange of resources – mobilization of local resources (economic, environmental, social), highlighting of amenities
	The sharing and creation of value for other territorial actors – the role of the VSE/SME in local economic development
	Relationship with the actors of the territorial field – professional, territorial (political) integration
	Exchanges through networks (partners, competitors, institutions, etc.) and increased confidence
	Loyalty and respect for exchanges with SMEs and local producers
	Percentage of local and regional products listed in sales outlets, including supermarkets
	The role and weight of territorial companies in local employment
Importance of history, values (rootedness and living together)	The commitment of producers, VSEs/SMEs to defend the values linked to the terroir
	History of the SME, of the distributor, their length of presence on the territory and their attachment to this space
	Tradition with regard to localized products (recipes, know-how, authenticity, etc.)
	Mobilization of local resources in terms of image, history, tradition, symbol, myth, legends and so on
	Action related to the search for meaning (societal concerns)
	Highlighting relationships with other local actors, facilitating consumer access to local products
	Promoting local products, the importance of local and proximity

**Analysis of semi-structured interviews with 29 consumers of localized products*

Table 6.8. *Characteristics related to territorial anchoring*

The consumers' discourse is analyzed as an act of committed and responsible consumption to support the territorial economy, by bringing their support to the actors of the territory. The textual analysis, in particular the classification carried out on the basis of the classified units of the corpus (88%, i.e. a high level of relevance), highlights the importance of territorial anchoring (class represents 10.51% of the classified textual units). The characteristics identified are of two kinds (Table 6.8):

(1) the importance of the links established between territorial actors (networks, partners and socio-economic relations) and (2) the attachment to the territory through the products promoted by the actors, allowing them to tie in with the history and values of this space and to share them.

In the end, territorial anchoring is also manifested through the process of innovation, transmission through mobility and acculturation. As a result, the terroir image of local products, like food cultures, is evolving and permanent, since both are co-constructed and shared by all actors in the territorial arena (Dedeire and Tozanli 2007). We can then think that territorial anchoring would promote better living together. Let us now look at the link between the terroir image of the local product and the factors of consumption value perceived by consumers, which could promote well-being and "eating together".

6.3.1.2. *The conviviality of the terroir: the links between place and body through local food*

The link between food consumption and health is an age-old issue for Hubert (2001). The literature shows that a healthy diet contributes to the achievement of central objectives such as health and well-being. Now a societal issue, better living and well-being are becoming increasingly important in our modern lives. Its growing interest is recent and is becoming one of the central concerns of marketing. Perceived holistically, the pursuit and development of wellness would be associated with many individual and societal benefits (Guevarra and Howell 2014). Food, sensitive to variations in the context (trends, crises, especially health crises), which influence the well-being versus the malaise of the consum'actor, impacts their health and ultimately, their better life.

The act of food consumption responds to the principle of incorporation (Fischler 1990), and eating constitutes a strong identity: "I am what I eat". Thus, by absorbing the food of the territory, the eater carries out an intimate encounter with the local cultures that they "incorporate" physically and symbolically (Fischler 1990). Food is seen as a vehicle for self-discovery: the search for experience can be likened to a journey organized around a quest for the self (Bessière *et al.* 2016), during which the eater enriches themselves, symbolically nourishing themselves (Fischler 1990). The functional or utilitarian value of a food is the extent to which a product has desired characteristics (attribute evaluation), is useful, and performs an expected function (Smith and Colgate 2007). Local terroir products are the assurance for the consumer of a superior quality guaranteed by the origin of the product, by the identification of the producer.

Beyond a physiological (functional) function, consuming local and terroir can have an affective dimension, triggering emotions linked to an attachment to the territory of origin and/or to a set of representations, symbols that the consumer associates with the product and the territory in which it is produced. The act of eating also has an imaginary function (Bessière *et al.* 2016); it carries memories of idealized images from the past. The act of buying and consuming is then guided by a feeling of nostalgia, a desire of returning to the roots (Aurier *et al.* 2004). The decision to consume is also driven by the search for the social value of products. The dimension of social commitment sheds light on the consumer's choice in relation to socio-political issues, or belonging to a group. The act of eating through the principle of incorporation (Fischler 1990) refers to a discovery and appropriation of the culture of the other. The development of social ties with territorial actors through local eating appears as an intercultural exchange and a desire to reconnect with one's roots, the peasant world (Bessière *et al.* 2016).

The research question is then the following: what is the influence of the terroir image of localized products on the consumption value perceived by the eater, and what is the impact on eating together? To answer this question, the theoretical framework used is the concept of the terroir image of the product developed above and the concept of the consumption value based on the experiential approach of Holbrook and Hirschman (1982), applied by Aurier *et al.* (2004) and by Smith and Colgate (2007). The fieldwork is based on a quantitative study (section 6.5, Box 6.2) conducted with 1034 consumers across the country.

6.3.1.2.1. Conceptual approach

The terroir image of localized products – as we have seen previously – is based on the notions of a situated and delimited territory (the terroir), and of proximity. The terroir is a construct, the result of practices and representations of territorial actors. Local terroir products are sought after for their qualities linked to the territory of origin. For the product to exist, three conditions must be met: (1) existence of a link with the origin, the place (geographical proximity), the strong links of the products and gastronomic recipes to the territory; (2) existence of a historical anchorage, the links to real or invented history; and (3) existence of a shared collective know-how, the links with the territory's actors. The perception of the terroir character of the local product is organized around the following two factors (Table 6.4):

– The "origin-craft-culture" factor associating the territory, region, land, as well as know-how, recipe, tradition and ritual.

– The "time and naturalness" factor associating history, nature and well-being.

The perceived value of consumption (section 6.5, Table 6.12) is apprehended from the experiential approach based on the production of value through the search for experience (Holbrook and Hirschman 1982). The integrative model from Aurier *et al.* (2004) and Smith and Colgate (2007) appears as an aggregate multidimensional construct of functional/instrumental value, hedonic/experiential value, symbolic/self-expressive value and sharing value (Table 6.9).

Facets of perceived value	Corresponding dimensions	Elements "origin" evoked, on which the value is based	Objectives of food wellness
Functional/instrumental	Utility, knowledge, quality, safety	Desired characteristics, place of origin, manufacturing method, cultural habits of the place of origin	Achieve health goals vs. avoiding risks
Experiential/hedonic	Experiential, hedonic, sensory, affective, social and epistemic stimulation	Exoticism, creation of sensory experience, feelings and emotions, artisanal character, authenticity	Promote taste pleasure and well-being in the consumption experience
Symbolic/expression	Self-expression, spirituality, integrity, responsibility and sharing	Prestige/status, community identity (ethnic consumption), ethical considerations, social, societal and environmental responsibility, citizen consumption, politics	Create social bonds in the sharing of meals to maintain a social norm and influence self- and group-identity

Table 6.9. *Selected facets and dimensions of consumption value and dietary well-being (Sources: Smith and Colgate (2007), Gabriel and Urien (2006))*

Of these three functions of food consumption, the hedonic function seems to play a role in the eater's well-being, and this is generally demonstrated by Guevarra and

Howell (2014). The consumption experience is then perceived as a desire to indulge (Holbrook and Hirschman 1982), even more so when well-being is associated with commensality (Fischler 1990). Guevarra and Howell (2014) indicate that sharing the experience would increase well-being, while enhancing a sense of group identity and affinity. Eating local and terroir is about eating together and sharing. It is a way to stay connected with others, close or not. Beyond the nutritional recommendations that have a role in being healthy, it is also about other determinants such as cultural, economic, environmental and especially social. Sharing the consumption experience and conviviality contribute to the quality of life and health.

6.3.1.2.2. Methodology and results of the study

The objective is to verify the existence of positive and significant links between the territorial origin of products and the variables sought through the consumption of this type of product. The methodology adopted combines exploratory factor analysis (EFA) and combinatorial factor analysis (structural equations), as well as multiple linear regressions. The exploratory factorial analyses test the reliability of the scales through EFA. The confirmatory factor analyses test the fit of the scale to the data, and then the structural model and linear regressions measure the links between constructs and test the research hypotheses.

Relationship tested	Coeff. β	p	Validation
IMAGT ↔ PCV	0.884	< 0.001	Validated
OPC – EXPE	0.049	0.427	Not validated
OPC – SYMB	-0.082	0.219	Not validated
OPC – FONC	0.783	<0.001	Validated
OPC – ETH	0.526	<0.001	Validated
TN – EXPE	0.555	<0.001	Validated
TN – SYMB	0.543	<0.001	Validated
TN – FONC	-0.028	0.578	Not validated
TN – ETH	0.000	0.997	Not validated

Table 6.10. *Validation of the links between IMAGT and PCV by structural analyses*

The results of the overall theoretical model test show the goodness of fit of the model to the data; the fit indices are in line with the literature recommendations. As a reminder (see section 6.2.2.2), the terroir image of products (IMAGT) is composed of two factors: origin–profession–culture (OPC), and time and naturalness (TN).

With regard to perceived consumer value (PCV), three facets were identified: functional (FONC), symbolic (SYMB) and experiential (EXPE), in accordance with the work of Smith and Colgate (2007), to which we must add ethics (ETH). The theoretical model is field tested (N=1034 respondents). The results of the test (Table 6.10) conducted by structural analysis show that the relationship between perceived terroir image and consumption value is significant (β=0.884, p<0.001). The relationship is validated. The influence of the terroir image on perceived value and its reciprocity confirms the work of Parissier and Langlois (2010) on the associations between types of perceived value and terroir products in a Quebec context, or the research of Merle *et al.* (2016), which shows the positive impact of local origin on food perceptions. Products, objects and events that are psychologically close and part of the individual's direct experience are perceived in a more concrete way than those that are distant (Merle *et al.* 2016), which could explain the increasing popularity of local products among consumers.

The indication of the origin of a food product, valued and interpreted as a sign of quality and safety by consumers, is a factor of value creation (Parissier and Langlois 2010). The smaller the territory, the greater the effect on value. Let us now look at the influence of each of the two dimensions of the terroir image (OPC and TN) on the four components of PCV (Table 6.10).

The functional value of the product is associated with the origin and know-how, according to the literature (Gabriel and Urien 2006; Parissier and Langlois 2010). This association is similar to the search for specific benefits (quality, taste, health/safety, etc.) to the original product through human factors, tradition and history.

The ethical value refers to individual and personal benefits outside the product itself. In the model tested, it is rather associated with the "origin-craft-culture" variable of the terroir image of the product. This type of association is highlighted by the work of Gabriel and Urien (2006) and Merle *et al.* (2016). The processes and conditions of production and/or manufacture linked to origin and proximity have a particular resonance with the respondent in terms of social and environmental responsibility, sharing and citizen consumption.

The experiential value, which corresponds to the hedonic function of consumption, is positively and significantly related to the "time and naturalness" variable. Pleasure, derived from the experience of consuming the product itself, is mainly activated by the criteria of conviviality and affective or social ties during consumption, in agreement with Parissier and Langlois (2010).

The link between the symbolic value and the "time-naturalness" component of the terroir image is positive and significant. The conditions of production and manufacture, the history of the product and of the people who have shaped it activate what Aurier *et al.* (2004) call "*the cause or values to which the individual aspires*". More generally, symbolic value refers to the psychological, imaginary and social significance (Smith and Colgate 2007) that the respondent attaches to the consumption of the product (status, prestige, self-image, community identity) (Gabriel and Urien 2006).

These results are correlated with the work of Merle *et al.* (2016) relating to local products on geographical location based on the strength of local identity, close to the regional belonging of Charton-Vachet and Lombart (2015). They show that the product of terroir origin, like the local product is perceived as better to health, better to taste. It provides better economic support to producers, is more respectful of the environment and, in the end, the purchase intention is higher. The local origin thus presents a double specificity: on the one hand, a small spatial distance between the place of consumption (consumer) and the place of production/manufacture, and on the other hand, the geographical belonging of the consumer to the territory of reference. The reduction of spatial and psychological distance to the product leads the consumer to imagine how the product was produced, manufactured and even distributed more concretely (Merle *et al.* 2016), which positively influences the perception of benefits for oneself (taste and health) and for others (social, economic and environmental). More broadly, the results related to local and terroir consumption show that beyond the undeniably present and important nutritional aspect, what happens upstream of the meal and around the plate are just as important for eating better and eating together. A shared meal would amplify sensory perceptions, with a real impact on well-being. Consuming local food meets the need for security, transparency and an appetite for local and regional products in what they carry. Eating together shows the will of the relationship, it is also a place of expression and acceptance of diversity, of affirmation of its tastes and opinions. We can thus argue that consuming locally has an effect on eating together (conviviality, the pleasure of sharing the consumption experience) and on living together better (symbolic and ethical facets: sharing, commitment and responsibility).

6.3.2. *Digital technology and better living together through local consumption*

Eating local and terroir food means becoming attached to the land, and reestablishing social ties by sharing more than just a convivial space. It is also to consume more transparently and responsibly, it is to be an actor in the co-construction of the territory for better living together. Today, digital technology is omnipresent and facilitates everyone's life. Consumers are transforming their lifestyles by adopting new digital technologies in their daily lives (mobile applications, connected objects, etc.), which facilitate the customer experience. In the age of digitalization, the consumer is on the move now more than ever, a term that reflects the mobility of the *shopper*, moving from one *device* to another, from one point of contact to another during the same journey, but not exclusively. Although the digital experience allows interaction with multiple interactive stimuli in the customer journey, it is also a co-constructed experience that allows the consumer to become not only an expert customer (*prosumer*), but also an eater-actor connected to others. The web can encourage connected local eating. Paradoxically, the use of digital technology raises questions about data transparency, but not only.

Is eating together (local conviviality vs. the loss of conviviality in today's food system) and better living together (duality between producers/polluters and quality food producers; short circuits and global warming; short circuits, transparency and health; local economy, sharing and solidarity) a utopia or is it a change in progress? Is eating local with the contribution of digital technology (its pluses and minuses) feasible and sustainable in the sense of sustainability? Eating locally is not reconnecting with others, and reweaving social links with local actors! Are the local eater and the connected consumer compatible?

To answer these questions, it seems necessary to take a holistic view based on the literature. First, we will look at the contributions of digital technology to the renaissance of local products and local commerce and to eating better, particularly through the development of applications. Then, in a second step, we will try to develop ways of thinking about the duality of the "local eater" (anchored territorially) and connected consumer (open to the world, to others).

6.3.2.1. *Eating local and digital: terroir and local connected*

The Covid-19 pandemic and lockdown have accelerated the consumption of localized food products (C-Way study, June 2020)[19], a trend in recent years, as has the digital transition[20] of small- and medium-sized businesses that has started. Local

19. A C-Way study estimates +113% for local produce delivery and +331% for choosing a local producer during the lockdown period. See: http://c-ways.com/wp-content/uploads/2020/06/C-Ways_Trendshaker_062020.pdf.

20. Google, for example, cites a study by the *Connected Commerce Council*, which found that nearly one in three small business owners believe they would have had to close their doors, at

commerce seems to be reborn thanks to digital technology. The development of these technologies makes it possible to access multiple points of contact at any time and in any place (Rogeon *et al.* 2019). Digitalization allows for the revitalization of the territory and the enhancement of local commerce by multiplying exchanges between consumers and merchants, and also directly with producers, according to the research work of Rogeon *et al.* (2019)[21], conducted on the basis of a literature review. Digitization is *a priori*, or at least often wrongly perceived, as having little impact on global warming. The example of Fraichy[22], online or via an app, allows people to order fresh products from small neighborhood shops online; furthermore, it provides home delivery by bike, cargo bike or car. Eating local and connected, beyond reinforcing the idea of the reterritorialization of our food, gives life and hope to small businesses, producers, craftsmen, farmers and SMEs that rely on their regions, territories, good products and traditional know-how, respecting the needs and expectations of consumers.

6.3.2.1.1. Local and terroir connected

Marketers have dubbed "Shopper 5.0", the new ultra-connected consumer profile, where the smartphone is almost an extension of the body. Their phone has become a personal assistant in various daily tasks. According to the Credoc digital barometer, in 2018, 89% of the French population were Internet users. Cell phone equipment reaches 94% or more than 45 million people.

This hyper-connected consumer easily navigates between *online* and *offline* purchases. They seek to use the different sales channels in combination, switching between them to cobble together a tailored hybrid experience (Rogeon *et al.* 2019). Producers are questioning the place to be given to digitalization in the marketing of their products (Rogeon *et al.* 2019), although relationships with the consumer are enriched by the spatio-temporal possibilities of Internet shopping – possibilities to interact with a retailer or other consumers on different touch points, anywhere and at any time.

The Covid-19 crisis has contributed to the acceleration of online shopping, which is perceived as more secure due to the absence of physical contact. Consumers believe that it has indeed allowed them to have access to essential

least in part, without digital tools (CB News). See: https://www.cbnews.fr/digital/image-google-ameliore-ses-outils-commerces-proximite-52379.

21. Their problem: "Understand the meanings associated (by consumers) with shopping for local products and their compatibility with those they associate with the digitalization of retail".

22. See: https://fraichy.com/.

products (for 31% of them), to bear the constraints related to lockdown (32.3%) and to avoid exposure to health risks (51.4%)[23].

The Internet has become a space that facilitates local relationships, local exchanges and human ties. The Internet makes it easy to get in touch with one's neighbors (*le bon coin* or *ShareVoisins*), to find a carpool (*Blablacar*) or to find a place to stay (Airbnb). The digital revolution is also at the service of new relationships between agricultural producers and individuals (*La Ruche Qui Dit Oui*, *marchand des quatre-saisons*) and a new online market has developed between producers and restaurant operators (*Cultures Chefs*)[24]. Consumers can reconnect with local producers and taste the products of their region.

Long before the Covid-19 crisis, direct relationships between consumers and producers had multiplied. Whether it is the AMAPs, the short circuits or the alternative distribution such as *La Ruche Qui Dit Oui* or the cooperative stores, many innovations were born to bring an answer to consumers in search of links with the origin of the consumed products. In addition, many producers have engaged in the development of Internet connections to get closer to their consumers and to involve their customers in a process of co-production of the consumption experience, for example in the choice of vegetables to be produced and the distribution methods. The Covid-19 crisis has put food and cooking back in the spotlight. A survey conducted by a team of French researchers on the basis of declarations collected in the NutriNet-Santé cohort, published online at the beginning of June on the MedRxiv[25] server, has shown that for a part of the French population, the lockdown has been an opportunity to spend more time in the kitchen, to make more purchases in local circuits and to eat more balanced meals. During lockdown, a slight increase in food procurement was observed for purchases on the Internet, by telephone or by mail order and for the purchase of food baskets from farmers or AMAPs.

6.3.2.1.2. Local products and local trade, a renaissance thanks to digital technology

Eating better is a need, and digital technology is a way to achieve it. A number of applications (Table 6.11) and initiatives other than social networks are being developed here and there, allowing the connected consumer to reconnect with local food and access products from their region more easily. This digital revolution is at the service of better local eating, in a more transparent way, and of giving one's

23. See: https://www.ecommercemag.fr/Thematique/retail-1220/Breves/Les-tendances-matiere-achats-internet-349871.htm, accessed June 12, 2019.

24. See: https://www.viaterroirs.com/.

25. See: https://www.medrxiv.org/content/10.1101/2020.06.04.20121855v1.full.pdf+html.

opinion and sharing it thanks to eWOM[26], electronic word of mouth, thus offering a rapid diffusion. This viral practice of spontaneous origin, which increasingly goes through the Internet and social networks, has a significant impact on the behavior of connected consumers.

Applications/sites	Features
Etiquetteable (collaborative)	Small encyclopedia of products, recipes and restaurants involved
Yes We Green	Directory of good plans to buy local (restaurants, wine shops, grocery stores, farmers' markets, AMAP, etc.), including clothing
La Ruche Qui Dit Oui	From the producer to the consumer: order, payment in a few clicks before collecting your order from a point near you
Potager City (bought by Carrefour Jan 2020)	Seasonal fruits and vegetables, subscription allows you to receive a mix of local products at a pick-up point near your home
Quitoque	Effortless recipes from French products ordered and received at home (refrigerated packages)
Péligourmet	The best of local products, operates like BlaBlaCar
Mooveat (collaborative)	Lists restaurants, suppliers, associations (AMAP, producers, etc.) within a 30-km radius of the selected address
Bioacasa	Delivery of baskets without moving (workplace or home)
Potagercity	Gateway to better, fresher and more responsible eating
Distri	Platform dedicated to short circuits, local products, local economy
Bioccinelle (Montpellier)	Allows the connection of local and organic consumers and producers – and craftsmen of Occitania
Tous Occitariens	Occitania Region platform to eat 100% local, 100% friendly (Sud de France brand ad)
Fraichy	Platform that allows you to shop online at local merchants in Montpellier and Toulouse
Mangez bien manger local	Lozère department website
Bienvenue à la ferme	Farm products and welcome at the farm
Localizz (13)	Online sales platform for quality local products at the right price
Lesamapdeprovence	Lists the points of sale in Provence

Table 6.11. *Examples of applications to eat healthy, local and seasonal food in a few clicks (not an exhaustive list)*

26. WOM: Word of Mouth – phenomenon of positive vs. negative recommendation of a product, service, brand, company within a network of acquaintances.

Beyond the purchase and consumption of local products in a few clicks, Holt (2007) proposes to encourage the development of decentralized electronic networks and platforms, which could support niche markets offering local products, and thus meet a growing demand. A strategy could be implemented that would allow a local supply network to incorporate local character elements (product attributes and shared meanings) that distinguish them from conventional or mass-produced products. An e-supply chain could integrate local actors, offering traceability, transparency, linkages, and any other information that responds to consum'actors concerns. This is the holistic approach developed by local food systems, a key driver of food sustainability according to Holt (2007), provided that a clearer definition of local food is adopted, which includes the attributes and shared meanings of local products and local eating, and also elements of connectivity that highlight the importance of human and social qualities of the territorial field. For example, the blockchain system, with the technology of NFC chips and QR code, allows the consumer to go towards more transparency by accessing the product's history and verifying the authenticity of their purchase.

To answer the paradox formulated above "can we be connected to the web and to our territory at the same time?" The answer is yes, albeit with some nuances. If digital technology ultimately offers the means to eat local together in complete transparency and trust, and to live better together – many arguments plead in their favor: better identification of local food, more pragmatic consumers, more autonomy and more pleasure, favoring the renewal of ties – however, there is a risk of dependency, a cost in terms of learning efforts and many other risks raised and/or relayed by Rogeon *et al.* (2019), including the fear of misuse of personal data or the reduction of the consumer's attention (the fact that they are doing something else at the same time while surfing). A loss of trust can even be observed when organizations are perceived as being non-transparent with regard to the use of their data (Portes, N'Goala and Cases, 2020)[27]. Despite these fears and risks, the initiatives developed show that we can be connected to the Web and to one's territory. The Web would allow both a gain and a loss of proximity, but concrete demonstrations remain to be made. Similarly, the literature emphasizes the positive contribution of digital technology to the customer experience, but to our knowledge, it does not address the risk of deteriorating the customer experience.

27. Digital transparency: dimensions, antecedents, and implications for customer relationship quality. To be published in *Recherche et Applications en Marketing*. See: https://doi.org/10.1177/0767370120935734.

6.3.2.2. *A consumer rooted in the territory and connected to others*

There are always good reasons to eat locally and be a connected consumer: for oneself (one's health, well-being, taste buds, wallet, commitments, etc.) and for others attached to the territory. Local issues such as ecology, resource preservation, agricultural practices, resilience to climate change, changes due to repeated food crises and pandemics, as well as social, economic and territorial food governance issues are sources of motivation, and also of concern.

According to Poulot (2014), these concerns lead to a strong demand for "eating well", which includes both "healthy eating" and "fair eating", one of the variations of which is "eating locally" (Fumey 2010). *Short food supply chains* in particular are favored as an alternative to the *globalized agrifood system*, contributing to the creation of social links between consumers and producers. The AMAPs are an example of this new order, also promoting exchanges between consumers. Would food[28] be a new stake in the development of territories to live better together?

Crises are disruptors, highlighting dysfunctions that aggravate imbalances and tensions, and stimulators of changes already at work, playing the role of catalysts that cause inflection points and ruptures. This situation affects the organization and functioning of supply chains that take food products from the producer to the consumer's plate. The problems of milk, fruit and vegetables highlight the fragility of vertical global systems (highly dependent and indebted producers who commit suicide, the increase in debt, the trade war between distribution giants, etc.). Consum'actors have become aware of the difficulty of sourcing, eating healthy and fair food in their own region and are changing their purchasing and consumption behaviors.

As we can see, crises are a powerful stimulator of food transition. Most distributors are committed to prioritizing local, available and seasonal production, and simultaneously, there is a very strong increase in demand for the direct sales networks that have already been established (Tardieu, Colibris, April 2020). At the same time, the crisis seems to be a formidable stimulator of digital transformation (LSA April 27, 2020). Internally, it allows employees to connect with each other, and it is also a stimulator of the quest for meaning for organizations seeking to establish a sincere and deeper relationship with their consumers.

---

28. L'alimentation, un nouvel enjeu de développement durable des territoires, Ministère de la transition écologique et solidaire, September 2017.

However, supply chains lack resilience – the ability to bounce back from stress like a rubber band that returns to its normal shape after being stretched. So how do we rewire the system? One solution lies in the ability to network on a territorial scale. Of course, developing short circuits, but above all rethinking the territorial economy by integrating the aspects of solidarity and circularity demanded by the consumer citizens. Producers, local VSEs and SMEs are able to bounce back from strong turbulence by including an element of horizontal collaboration (with others: sharing and responsibility) to vertically structured organizations. As networks of producers and consumers grow, they become more connected and are more able to adapt flexibly to meet local demand, creating a territorial mosaic that can withstand the turbulence of the dominant vertical system. Economists show that it is the diversity of food consumption patterns in Europe that generates the added value of this major economic sector (Daniel 2020)[29]: move towards flexible and diversified food systems that reflect our territories!

Food, a socially and spatially constructed fact, is a highly cultural and geographic product for Fumey (2010). Geography – in terms of territory, localized system, origin, relocation of supplies and so on – once again becomes an instrument of understanding for the local eater. It allows them to remain an actor in their food; in other words, it provides them with the necessary knowledge to give meaning to what they eat (Fumey 2010) and to share it with others. Consumers are now, more than ever demanding transparent, respectful and meaningful communication. There is a change in habits and behaviors boosted by the global health crisis of 2020, like most crises. As proof, 60% of marketers will increase their e-commerce budget in the next months and years according to a study conducted during the last two weeks of April 2020 by Dentsu Aegis Network, among 701 companies in 36 countries. This is to meet the strong expectations of consumers in terms of social responsibility, and fluidity in the user experience. For advertisers, communication must become more responsible in response to consum'actors expectations. In this context, the mix of "digital and physical presence" is becoming increasingly important and is pushing advertisers to digitalize their communication and distribution.

Do concerns about "consuming locally" and "eating better" herald a new pact between producers and consum'actors? A number of French regions (Brittany, Grand-Est, etc.) have initiated regional food projects: the "Breizh Alim" initiative in Brittany and the "Assises alimentaires de la région Grand Est" in Grand-Est. These actions, led by the regions, aim to organize meetings between local buyers and suppliers, in order to develop the consumption of local products and *ultimately* respond to the concerns of citizens. In the United States, a country not known for its

29. Director of research and economist at the ESA in Angers, April 6, 2020 in Ouest-France.

food tradition, initiatives are emerging, notably in New Jersey[30], where the fourth smallest farms in the United States are located. Producer and consumer networks are becoming more connected and are organizing themselves to produce, distribute and consume locally and *ultimately* live better together!

6.4. Conclusion

Today's dominant food systems, which are mainly based on agroindustry – an industrial, specialized and hyper-concentrated agriculture – are strongly criticized for their negative externalities on the environment, the development of territories, employment and human and non-human well-being, and on the abiotic environment. Localized products (terroir products, local products, products from here, consumed here, etc.) are an alternative to the dominant model and could allow people to eat better (health) and live better within their territory, by promoting the reconnection between food and local production. More and more consum'actors are in favor of food relocalization, a phenomenon benefiting from the accelerating effect of crises such as the Covid-19 crisis. Eating locally – more than one French person out of four consumes or associates local products to their food – is a practice that is gaining momentum. Digitalization that is well underway should facilitate the reconnection to this food transition desired by a growing number of consum'actors. The territorial anchorage and proximity sought, the terroir image of local products are factors of trust, conviviality and sharing with other territorial actors. Digital technology contributes to buying local, and even to eating connected with others, although there are some obstacles. Eating locally and being a connected consumer are compatible. However, many points need to be improved, including the scope of the local product (is it that important?), the price (a major obstacle), the difficulty in obtaining supplies, the paradox of the digitalized eater who is both close to and far from "social reality", but not exclusively. It remains to be seen which levers will allow the relocalization of food systems to be implemented in a sustainable way. The final word is borrowed from Paddeu (2017), for whom it seems essential to build alliances between humans and non-humans, in order to contribute to "the reinvention of eating locally, not as a moralizing watchword, but as the *de facto* sharing of a common territory, where alliances between human and non-human collectives make good politics".

---

30. See: https://theconversation.com/new-jerseys-small-networked-dairy-farms-are-a-model-for-a-more-resilient-food-system-137881, accessed June, 122020.

6.5. Appendices

Motivations to consume local and terroir products were studied through a mixed methodology, combining a qualitative study in two cohorts of consumers of local and/or terroir products, and a quantitative study corresponding to three data collections from consumers.

Qualitative study

The qualitative study included two cohorts of 16 and 13 semi-structured interviews with consumers of local products, selected from sales outlets (supermarkets and small local merchants) and knowledge networks. The interview guide used made it possible to address the consumption of localized and/or local products (choice, consumption, etc.) and, more broadly, the relations of such consumption with the territory and local actors. Analysis of the textual data of the fully transcribed speeches was carried out using Alceste software. Analysis of the textual data allows the emergence of the main criteria related to the purchase and consumption of local products and their anchoring to the territory of belonging.

Quantitative study

The field study was extended by three quantitative surveys. A first data collection gathered 403 valid questionnaires from consumers spread over the national territory, dissemination takes place through a snowball effect from the researchers' own networks via the Internet and covers almost the whole of France. A second collection gathers data from 279 respondents of local and terroir products from the South-East of France. It was followed three months later by a third collection of 352 respondents from the whole country and carried out from a panel. The quantitative data was analyzed using IBM-SPSS and Amos software. The models used allow us to test the influence of the terroir image on trust and perceived value. The last two surveys feed the exploratory analysis (Principal Component Analysis – PCA) and the confirmatory analysis of the construction of measurement scales from structural equations and allow the testing of the network of relations regulating the conceptual framework:

- The objective of the exploratory factor analysis of the PCA type was to purify the measurement scales and study their dimensionality, to assess their reliability (factorial contributions > 0.7 with $t > 1.96$ and Jöreskog's rhô > 0.7) and their convergent validity (rho vc > 0.5).

- The confirmatory factorial analysis conducted on the basis of structural equations helped to validate the psychometric qualities of the measurement scales (reliability and convergent and discriminant validity), and with the regression analyses (simple and multiple), allow the measurement of the links between the constructs.

Box 6.2. *Methodology*

Perceived terroir image of products

Indications of origin (official or unofficial labels) and appellations of origin are associated with a territorial location, a particular know-how, and a perceived authenticity that cannot be relocated, and are therefore anchored in a limited and well-identified area. The scale used to measure the terroir image by Aurier *et al.* (2004), based on perceived sources of terroir (origin and profession; time, culture and nature), was adopted and adapted.

Consumer confidence

Trust promotes interaction between partners and reduces uncertainty (Gurviez and Korchia 2002). For Morgan and Hunt (1994), trust is seen as a set of activities that enable the establishment, development and maintenance of relational exchanges. As this concept is the subject of a relative consensus among researchers, we will retain its three-faceted approach, integrity, credibility and benevolence, the combination of which determines the level of trust of a consumer towards the brand (Gurviez and Korchia 2002). Kaâbachi (2007) shows that trust in the organization reflects the belief that the organization is credible, benevolent and capable of defending the interests of its customers and stakeholders.

Perceived consumption value

The consumption value is linked to the person-object experience, highlighting the experiential and symbolic dimensions in addition to the functional dimension of the product. Analysis of the consumer's point of view of the consumption value of fruits and vegetables in the work of Gurviez and Sirieix (2010) reveals three facets of PCV: hedonic, instrumental and self-expression. Ambroise, Brignier and Mathews (2010) base their work on the perceived value of private label with three dimensions: utilitarian, experiential and symbolic. These three common facets constitute the value perceived by the consumer, to which we have added the ethical dimension in the purchase and consumption of localized products, operationalized in five adapted statements.

Box 6.3. *The mobilized constructs*

People interviewed	No. of interviews/ No. of words	Characteristics interviewed
Consumers	13/40,392	Knowledge networks: 51.2% women; average age 41.6 years (26 to 66 years); six PCS* represented (46.2% executives); 76.9% with two or more years of higher education; 53.8% urban origin.
Consumers	16/35,614	Selection at the exit of points of sale – interviews in the homes of residents: 56.2% women; average age 42.5 years (23 to 74 years); eight PCS* represented (37.5% executives); 87.5% with two or more years of higher education; 68.8% urban origin.

*PCS: Professions and Socioprofessional Categories

Table 6.12. *Presentation of the qualitative fieldwork*

Built	Authors	Dimensions, number of items
Terroir image (perceived sources)	Aurier, Fort and Sirieix (2004); Fort (2005) – adapted scale	Two-dimensional - Origin/business: nine items - Time/culture: nine items
Perceived consumption value	Parissier and Langlois (2010); Aurier *et al.* (2004) – adapted scale	Multidimensional - Functional: ten items - Experiential: five items - Symbolic: nine items - Ethics/sharing: five items
Trust	***Trust in the brand***: Gurviez and Korchia (2002)	- Credibility: three items - Integrity: three items - Caring: two items
	Trust in the label: Kaabachi (2005)	One dimensional: three items

Table 6.13. *Scales for measuring constructs*

Socio-demographic characteristics		N=631		N=403
Workforce	Number of consumers	279	352	403
Localized products	Most purchased product category	Delicatessen	Delicatessen	Cheese Wine
Gender	Male	45.5	56.0	49.1
	Female	54.5	44.0	50.9
Age and age groups	Average age	44.8	47.7	40.2
	25–49 years old	51.3	54.0	53.3
	50-64 years old	25.1	39.5	19.4
	> 65 years old	9.3	11.4	9.4
Level of education	Baccalaureate (Advanced Level) and higher *+4/5*	10.0	24.1	32.2
	Baccalaureate +2	26.9	20.7	21.6
	Baccalaureate	24.7	20.7	14.9
Place of residence/origin	City	60.9	60.5	58.8
	Countryside	39.1	39.5	41.2
	Urban	62.0	61.4	51.4
	Rural	38.0	38.6	48.6
Household size	Couple	74.2	75.0	70.0
	Couple + 0 children	43.0	36.9	34.7
	Couple + 1 child	26.9	24.4	20.6
	Couple + 2 children	16.1	16.4	14.7
Activity sector SPC	Professional executive intellect support	11.5	28.7	23.1
	Employee	31.5	19.9	23.1
	Retired	15.8	18.8	13.9
	Intermediate profession	11.5	15.3	5.7

Table 6.14. *Socio-demographic structure of the sample groups*

6.6. References

Abid, T., Rodier, F., Durif, F. (2020). Produits alimentaires locaux : les motivations d'achat en fonction des circuits de distribution. *Décisions Marketing*, 98.

Aprile, M.C., Caputo, V., Nayga Jr., M. (2016). Consumers' preferences and attitudes toward local food products. *Journal of Food Products Marketing*, 22(1), 19–42.

Arnould, E.J. and Thompson, C.J. (2005). Consumer culture theory (CCT): Twenty years of research. *Journal of Consumer Research*, 31(4), 868–882.

Arsil, P., Brindal, M., Sularso, K., Mulyani, A. (2018). Determinants of consumers' preferences for local food: A comparison study from urban and rural areas in Indonesia. *Journal of Business and Retail Management Research*, 13(2), 184–195.

Aurier, P., Fort, F., Sirieix, L. (2004). Les produits de terroir du point de vue des consommateurs : sources perçues et associations au terroir. *Actes du XXème Congrès St Malo, AFM*, 6–7 May, 1–32.

Bérard, L. and Marchenay, P. (2004). *Les produits de terroir. Entre cultures et règlements.* CNRS Editions, Paris.

Bessière, J., Mognard, E., Tibère, L. (2016). Tourisme et expérience alimentaire. Le cas du Sud-Ouest français. *Téoros. Revue de recherche en tourisme*, 35, 2 [Online]. Available at: http://journals.openedition.org/teoros/2928 [Accessed 18 June 2020].

Bousquet, F., Barbat, V., Petzold, S. (2020). La nature stratégique des ressources issues de l'ancrage territorial : proposition d'une grille d'évaluation pour les PME. *Revue Internationale PME*, 33(1), 41–73.

Bowen, S. and Mutersbaugh, T. (2014). Local or localized? Exploring the contributions of Franco-Mediterraneean Agrifood theory to alternative food research. *Agriculture and Human Values*, 31(2), 201–213.

Casabianca, F., Sylvander, B., Noël, Y., Béranger, C., Coulon, J.B., Giraud, G., Vincent, E. (2006). Terroir et typicité : propositions de définitions pour deux notions essentielles à l'appréhension des Indications Géographiques et du développement durable. *Actes du Vième Congrès International des Terroirs Viticoles*, 3–7.

Charton-Vachet, F. and Lombart, C. (2015). Nouvelle approche conceptuelle et opérationnelle du lien entre un individu et sa région : l'appartenance régionale. *Recherche et Applications en Marketing*, 30(1), 52–80.

Dedeire, M. and Tozanli, S. (2007). Les paradoxes des distances dans la construction des identités alimentaires par acculturation. *Anthropology of food*, S3 [Online]. Available at: http://journals.openedition.org/aof/2582 [Accessed 04 November 2019].

Delfosse, C. (2012). La France et ses terroirs. Un siècle de débats sur les produits et leurs liens à l'espace. *Revue POUR – GREP Groupe Ruralités Education et Politiques*, 215–216, 63–74.

DiMaggio, P.J. and Powell, W.W. (1983). The iron cage revisited: Institutional isomorphism and collective rationality in organizational fields. *American Sociological Review*, 48, 147–160.

Dion, D., Rémy, E., Sitz, L. (2010). Le sentiment régional comme levier d'action marketing. *Décisions Marketing*, 58(2), 15–26.

Dufeu, I. and Ferrandi, J.M. (2011). Proximité perçue, confiance, satisfaction et engagement des consommateurs dans le cadre d'une économie du lien : les AMAP. *5èmes Journées de recherches en sciences sociales*, Dijon, December.

Feldmann, C. and Hamm, U. (2015). Consumers' perceptions and preferences for local food: A review. *Food Quality and Preference*, 40(A), 152–164.

Fischler, C. (1990). *L'Homnivore*. Éditions Odile Jacob, Paris.

Fumey, G. (2010). *Manger Local, Manger Global*. CNRS Editions, Paris.

Gabriel, P. and Urien, B. (2006). Valeurs de consommation et origine territoriale des produits. *Décisions Marketing*, 43(44), 41–54.

Granovetter, M. (1985). Economic action and social structure: The problem of embeddedness. *American Journal of Sociology*, 91(3), 481–510.

Guevarra, D.A. and Howell, R.T. (2014). To have in order to do: Exploring the effects of consuming experiential products on well-being. *Journal of Consumer Psychology*, 25(1), 28–41.

Gurviez, P. and Korchia, M. (2002). Proposition d'une échelle de mesure multidimensionnelle de la confiance dans la marque. *Recherche et Applications Marketing*, 17(3), 41–59.

Holbrook, M.B. and Hirschman, E.C. (1982). The experiential aspects of consumption: Consumer fantasies, feelings and fun. *Journal of Consumer Research*, 9, 132–140.

Holt, G. (2007). Local food in European supply chains: Reconnection and electronic networks. *Anthropology of Food*, S2 [Online]. Available at: http://journals.openedition.org/aof/479 [Accessed 04 February 2020].

Hubert, A. (2001). Systèmes agroalimentaires localisés, réflexions d'une anthropologue. In *Systèmes agroalimentaires localisés : terroirs, savoir-faire, innovations. Etudes et recherche sur les systèmes agraires et du développement*, Moity-Maïzi, P., Geslin, P., Muchnik, J., Sainte Marie, C., Sautier, D. (eds). INRA, Paris.

Kaâbachi, S. (2007). La valeur perçue : une variable stratégique pour les enseignes de distribution alimentaire. *2ème journée du marketing IRIS, "la relation client dans les activités de service"*, Lyon, 15 March.

Kantar (2020). In Livre-Blanc SIAL Insights. SIAL, Paris.

Lenglet, F. and Müller, B. (2016). Les effets de la proximité sur la confiance : le cas des produits localisés. *32ème Congrès de l'Association Française du Marketing*, Lyon, 18–20 May.

Merle, A. and Piotrowski, M. (2012). Consommer des produits alimentaires locaux. Comment et pourquoi ? *Décisions Marketing*, 67, 37–48.

Merle, A., Herault-Fournier, C., Werle, C. (2016). Les effets de la mention d'origine géographique locale sur les perceptions alimentaires. *Recherche et applications en marketing*, 31(1), 28–45.

Morin, E. (1973). *Le paradigme perdu : la nature humaine*. Le Seuil, Paris.

NIELSEN (2019). In *Vigie Grande Distribution* (2020). Editions Dauvers, n°344.

Paddeu, F. (2017). Manger local. Leurres et promesses. *Association Vacarme*, 4(81), 40–45.

Parissier, C. and Langlois, M.-C. (2010). Valeur de consommation des produits agroalimentaires de terroir : une contribution à la mesure du concept au Québec. *Actes de la 6e journée AFM du marketing agroalimentaire de Montpellier*, October.

Pesqueux, Y. (2015). L'innovation entre tradition et nouveauté. *Vie et sciences de l'entreprise*, 2(200), 99–118.

Poulain, J.P. (2018). *Dictionnaire des cultures alimentaires*, 2nd edition. Presses Universitaires de France, Paris.

Poulot, M. (2014). *Histoires d'AMAP franciliennes : quand manger met le local dans tous ses états. Territoire en mouvement. Revue de géographie et aménagement* [Online]. Available at: http://journals.openedition.org/tem/2388 [Accessed 02 February 2020].

Quellier, F. (2018). THOMAS PARKER, Tasting French Terroir. The History of an Idea, Berkeley, University of California Press, 2015, 229. In *Revue d'histoire moderne contemporaine*, 65(2), 168–170.

Rogeon, F., Michaud-Trévinal, A., Collin-Lachaud, I. (2019). Le digital, la recherche du graal ? Mise en perspective de paradoxe entre la digitalisation du commerce de détail et le magasinage de produits alimentaires locaux. *Colloque Etienne Thil*, Paris, 22.

Scott, W.R. (1995). *Institutions and organizations*. Sage, Thousand Oaks, CA.

Smith, J.B. and Colgate, M. (2007). Customer value creation: A practical framework. *Journal of Marketing Theory and Practice*, 15(1), 7–23.

7

By Way of an Epilogue: "Eating Together" in the Time of Covid-19

Gilles SÉRÉ DE LANAUZE[1] and Guillaume LE BORGNE[2]

[1] *MRM Laboratory, University of Montpellier, France*

[2] *Institut de Recherche en Gestion et en Économie (IREGE), University of Savoie Mont Blanc, Chambéry, France*

7.1. Introduction

The Covid-19 crisis and the experiences of lockdown, from March to May 2020, and then the subsequent lifting of lockdown have dramatically altered the eating habits and behaviors of most people. Confined to their homes, with limited access to products and stores, without opportunities to meet their friends at home or in restaurants that had closed, people all over the world have ended up having to reinvent "eating together". Between constraints and opportunities, between the prohibition of meeting up and the rediscovery of home cooking, the practice of eating and commensality has taken on new forms and new relational modes. It is still too early to judge the long-term effects of these changes; however, they have certainly left their mark on the way we eat together. These changes have not only modified our daily food practices, but perhaps also our perceptions and beliefs about food consumption, production and distribution. This chapter, which unlike the previous ones is not based on the authors' own research findings, proposes a reflection prompted by the events and some avenues for future research. It presents these reflections around four major themes: the impacts of lockdown and the lifting of restrictions on food practices; the central role of digital technology and the

Evolution of Social Ties Around New Food Practices,
coordinated by Gilles SÉRÉ DE LANAUZE.

changes induced in terms of product sharing and accessibility; the changes in values, concerns and priorities resulting in a reinforcement of certain "responsible" diets and lifestyles; and finally, the effects on supply and the refocusing on the local, in terms of distribution, and also of production and services.

7.2. The change in practices

Following the announcement of lockdown, a first reaction was observed, quite symbolic of a perception of food as the first resource to be secured in times of crisis: the rush and the constitution of stocking up by many consumers[1], on many products such as basic cereals (pasta, rice), frozen products, water or long-life products. At the same time, there was a clear disaffection for Asian and Italian products[2], countries directly associated with the crisis in these first weeks of the pandemic. These behaviors, which were certainly short-lived – they became rarer afterwards – were the marker of a refocusing on food as a vital priority, and the revelation of food anxieties ready to emerge at each new health crisis.

The period of lockdown then provided an opportunity to redefine eating together. In some households, this permanent grouping in the same space greatly increased the number of meals eaten together, necessitating a rethinking of some of the usual practices, tasks and roles. For example, in households where only one of the two members had a job outside of the home, they often became responsible for the food supply because of their trips to work.

More planning was needed, especially to limit the time spent in stores and the number of trips to busy locations, and also because of the increase in the number of meals taken at home. As soon as purchases were planned, a deliberation on the desires and needs of each individual developed. For some households, more time was spent cooking (increase in the number of meals eaten at home), with the possibility of getting together to prepare meals, especially for households with children. The sharp increase in the purchase of basic products for recipes (flour, eggs, oil, cream, butter, etc.) reflected an increase in homemade food[3]. The need to

1. IRI-BCG survey, Covid-19 Impact, Consumer spending tracker for measured channels, U.S., UK, France, Italy, Germany, Spain, New Zealand, April 30, 2020. Available at: https://www.iriworldwide.com/IRI/media/Library/2020-06-05-IRI-BCG-COVID-Global-Consumer-Spend-Tracker.pdf.

2. This is probably due to the fear of contamination of these products by the virus. China was chronologically the first country in the world to be affected by the epidemic, and in Europe, it was Italy.

3. France Agrimer study, September 2020, L'impact de la crise de la COVID-19 sur la consommation alimentaire en France: parenthèse, accélérateur ou élément de rupture de tendances?

structure time was increased by the situation, and this could be done in particular with the help of meals, taken together, moments of coordination of very diverse individual agendas within the household. Beyond a more frequent and complete commensality, which is also sometimes more constrained, certain food planning and management practices have been implemented more frequently within households. This is evidenced by certain indicators that have increased, such as the more frequent use of a shopping list (+34%), checking stocks at home before making purchases (+47%), batch cooking[4] (+23%), creating recipes from available ingredients (+33%), and regular organization of refrigerator and cupboard storage (+36%)[5]. Often implemented out of necessity, they are nonetheless sustainable practices, in the sense that they lead to less waste[6].

On the contrary, for relatives, partners or more distant friends, this lockdown marked a physical distancing. The practice of aperitifs or meals between friends by videoconference has become widespread, with the aim of circumventing or reducing, at least symbolically, the distancing imposed. In these moments of food conviviality, some may have decided that the guests (if not commensals) would eat the same dishes (cooked in each home), for example the same birthday cake. The different versions of this dessert were then a subject of online discussion between guests, of sharing and socialization. Part of the symbolism of sharing the same food was thus preserved. Other groups met online without trying to eat the same thing, which did not prevent them from being interested in what the other had prepared, and from discussing it (preferences, tastes, feelings, exchanges of recipes, etc.). They ate at the same time, connected, but did not eat the same thing: here, the fact of not eating the same thing had no reason to stand out as an anomaly (since the situation made it "normal"). And the variety of the contents of the plates in these meals shared online did not refer the guests back to their differences, as could have been the case around the same physical table. But are these practices observed during lockdown, motivated essentially by the need to react to such a strange and unusual situation, likely to last? It is permissible to doubt this, given the increasing number of cases of people seeking to eat together at all costs, in the same physical space, even though it means ignoring the prohibitions.

Available at: https://www.franceagrimer.fr/Mediatheque/INFORMATIONS-ECONOMIQUES/TOUTES-FILIERES/ETUDES-ET-SYNTHESES/2020/L-impact-de-la-crise-de-la-COVID-19-sur-la-consommation-alimentaire-en-France-parenthese-accelerateur-ou-element-de-rupture-de-tendances.

4. Batch cooking: preparing all of the meals for the week at once.

5. See: https://wrap.org.uk/sites/files/wrap/Citizen_responses_to_the_Covid-19_lockdown_0.pdf.

6. See: https://theconversation.com/comment-le-confinement-peut-durablement-transformer-nos-pratiques-alimentaires-142205.

Lockdown should not be idealized as a period of purely improved "eating together": the grouped households witnessed the occurrence of frictions, particularly when the needs, desires or values of their members proved incompatible, and it became necessary to debate each other's food choices and make compromises. While some households cooked more fresh food, others resorted to long-shelf-life ready-made meals and frozen food, because of their convenience and the possibility of storing them. For some people, meals at home have allowed them to structure the time of a reconfigured daily life, while for others, the days have become unstructured and the food they eat has changed. Finally, while some have found meaning and reassurance in physically visiting local food shops, others have avoided such places and have relied massively on online orders and farm drives. In terms of conviviality, those who were physically separated were only partially able to bridge this distance through meals shared remotely and online. The drop in sales of festive alcohols (liqueurs and bubbly wines in particular) during lockdown, and especially their rebound following the May lifting of restrictions[7], testify, beyond seasonality, to a postponement of expressions of conviviality to physical reunions.

This lockdown situation thus sheds light on the dimensions of eating together that were accentuated by necessity (planning, preparation), or that the groups of individuals tried to reconstitute (temporal synchronization, similarity of dishes, socialization around the food eaten) in order to derive the usual symbolic, social and affective benefits and maintain a form of relational proximity. It also showed the diversity of the modes of adaptation, largely dependent on the contexts, constraints, skills and psychological reactions to the crisis of each one.

7.3. Irruption and trivialization of the digital in food and conviviality

The lockdown period has seen a marked rise in the use of digital technology. Many moments of sharing, exchanging, conviviality and information have shifted to remote modalities, such as the creation of online groups, chefs' blogs presenting tips and recipes, and of course, online ordering. Food e-commerce has strongly increased during the lockdown (up to +100% the week of April 6, 2020 compared to the same week in 2019[8]), especially for supermarkets and farmers (farm drives). Telecommuting – which is still encouraged after lockdown during which it became the norm – leads to more meals eaten alone at home and a less clear delimitation of meals, possibly taken in front of the screen. The absence of social constraints can lead to not taking a real break for the meal, and the disappearance of meals taken

7. France Agrimer study, ibid.

8. Source: Nielsen, May 28, 2020. Available at: https://www.nielsen.com/fr/fr/press-releases/2020/internet-relais-du-e-commerce-alimentaire-pendant-le-confinement/.

with colleagues leads to isolation and a reduction in the symbolic relaxation and pleasure that could (sometimes) be derived from them. With the lifting of restrictions, the widespread use of digital technology has continued in a context of ever-present sanitary measures. In addition to videoconferencing meals with friends or family and ordering online, the increase in the use of digital technology has led to new online activities such as consulting recipes, sharing food photos with friends and family, simply following the progress of one's diet, or participating in online communities around food (recipes, advice, exchanging photos and experiences). Periods of lockdown and lifting of lockdown are therefore part of the acceleration of a long-standing trend of a society that is digitizing almost all aspects of its daily life, with the cell phone as the main entry point: social networks, payment, entertainment, interaction with brands, stores and other consumers.

Digital technology certainly allows us to partially overcome physical distance to eat together, order a product, or visit the production line of a product virtually. Nevertheless, it has limits and proven defects, linked to the impossibility of totally overcoming the spatial and physical necessities and contingencies. First, virtually eating together, as mentioned above, does not equal physical commensality. Moreover, the hyperconnection of individuals can lead to a reduction in our eating experience. Indeed, hyperconnection reduces the awareness of the food act, when we eat in front of a screen or while receiving notifications. It also removes the contribution of the use of senses such as touch and smell, when shopping online and implies ensuring consumer trust by other means. Finally, if it allows a certain form of rationalization of consumption, it moves away from the feeling of needs, and disrupts a commensality that gives more room to a virtual elsewhere than to the commensals physically present.

In the *fast-moving consumer goods* (FMCG) sector, brands and retailers are the significant driving forces of this long-term evolution. First, digitalization is a source and a tool for real-time data processing, enabling the optimization of production and supply chains and the personalization of automated customer relations. It also allows for greater transparency on manufacturing processes[9], on the one hand, to compensate for the physical distance that it puts between the consumers and products, which the former can no longer touch when purchasing and, on the other hand, to regain the trust of consumers anxious about health scandals and the "black boxes" that packaging and their ultra-processed content have become. The use of blockchain also contributes to the optimization of supply chains, and promises better traceability and therefore better consumer confidence.

9. See, for example, Fleury Michon's #VenezVérifier application.

Finally, the optimization of supply chains that digital technology made possible, and that was desired by the distribution industry[10], showed its downside in a striking way during the Covid-19 crisis. These supply chains, which are particularly complex, long, interdependent and optimized, proved to be considerably vulnerable to the unexpected, sometimes creating widespread supply failures following the breakage of a single link. Thus, the announced optimization – real when everything is going well – has been marred by major systemic fragilities. In contrast, the resilient nature of local production-consumption has been masterfully asserted in recent months.

In addition, digital, which certainly allows the arrival of "newcomers", such as DNVB[11], benefits above all the big ones, because of their ability to collect more data than others and thus really benefit from the contributions of Big Data, while having an omnichannel presence. As Jean-Paul Agon of L'Oréal explained in 2019: "Contrary to the idea that circulates, digital benefits the big brands. Of course, it allows many small brands to appear but, in the end, the algorithms mean that it is the big ones that emerge". In food, big brands such as Danone or Nestlé certainly benefit from such mechanisms.

7.4. Strengthening communities and beliefs?

This health crisis has also reactivated broader concerns of a macro-economic, societal and environmental nature. Globalization, the sharing and protection of resources, biodiversity, growth, all of these issues have come back to the forefront of debates and have led to new injunctions in terms of food practices, sometimes for almost philosophical reasons, and advocating, for example, homemade, local and meatless food. In this respect, it is symptomatic that a Chinese food market, the one in Wuhan that made the headlines in March 2020, was presented as the origin of the pandemic. It did not take much more to reinforce questions about the links between food consumption and disease, and more broadly, the place of humans in the universe. For example, the association Végétarienne de France has put forward the hypothesis of possible links between the production and consumption of wild meat (and also from industrial farms) and the propagation of deadly viruses[12]. This community thus sees Covid-19 as an additional reason and illustration of the need to reinforce and continue its practices, and is not alone in expressing such concerns.

10. "We must rethink the consumption model by reducing waste and optimizing production and consumption channels", Laurent Gueye, Adoria.

11. "Digital Native Vertical Brands", a name for brands born online that master their entire value chain.

12. See: https://positivr.fr/coronavirus-la-tribune-de-lassociation-vegetarienne-de-france/.

Conversely, it is quite possible to envisage a decline in the impact of consumption as a vector of identity for some consumers, following the loss or reconfiguration of reference points during lockdown, and the multiplication of the challenges and injunctions perceived by the consumer. The need for conviviality following a relative deprivation could, with the multiplication of constraints and fears, contributes to a disengagement from the community in which they were involved for some. Will the underlying and structuring convictions of certain chosen communities resist the constraints of isolation and social risk? The communities thus appear to be both a place of safety and reassurance of identity, and also a potential brake on safety, practicality, or the pursuit of other economic objectives associated with food.

At the same time, many individuals, who are not members of any vegetarian or other food practice community, nor of any ecological group, have nevertheless modified part of their food practices during lockdown in the direction of greater parsimony and responsibility. The time to think, the time to make their own food instead of buying processed products, the constraints in terms of supply and the shortage of stocks of delocalized products, are among the factors that pushed them towards a more responsible food consumption. Conversely, however, other less sustainable practices have developed because of the crisis. Individual packaging and plastic could be considered as a bulwark against the virus, including for fresh products, and overstocking[13] appeared to be an adequate response to the perceived risk. What will be the outcome when the crisis is over? How can sustainable practices be maintained by those who have newly implemented them? With the lifting of lockdown, it is still not possible to talk about a return to normalcy, but the context has changed again and there is no guarantee that old practices will not be reinstated. However, how can we avoid unsustainable practices adopted during lockdown from taking hold (due to a regression of the environment regarding individual priorities, to a situation considered exceptional, to organizational difficulties, etc.)? Organic products certainly progressed more strongly than conventional ones in the spring of 2020[14]. Yet, when the effects of the economic crisis are at their strongest, won't the economic constraints of some households be an insurmountable obstacle to the purchase of these more expensive products? Thus, the economic difficulties induced by the health crisis can both generate some more sustainable consumption patterns, with forms of sobriety, or even deconsumption, and also at the same time, a lesser appetite for products with lesser environmental

13. See: https://theconversation.com/comment-le-confinement-peut-durablement-transformer-nos-pratiques-alimentaires-142205.

14. See: https://www.nielsen.com/fr/fr/insights/article/2020/la-sante-du-bio-en-france-au-revelateur-du-covid-19/.

impact, in connection with concerns focused primarily on purchasing power and health, and economic imperatives. And even though, as during lockdown, consumers opt for more simplicity, will they choose home-made, raw products, with the idea of reappropriating their food, of voluntarily simplifying its content, of consuming soberly and aiming at a form of authenticity? Or, on the contrary, will they be looking for opportunities, practicality and security, and thus be tempted to favor home delivery of processed and over-packaged products?

7.5. A refocusing on the local and nearby

After an initial moment of panic that saw many consumers rush to buy basic necessities for fear of a shortage, the lockdown consumption pattern has largely turned to local and proximity forms, often in association with online ordering (and farm drives). Contrary to many other categories that have seen an explosion of globalized online trade, in food, consumers have gradually refocused on more local products and producers, combining, on the one hand, distrust of imports and distant origins (many of the latter were in fact stopped by border closures) and, on the other hand, a desire to defend local actors in times of employment and activity crisis (producers, restaurant owners, retailers). In addition, as the preparation of "good food" and "good products" took on a new importance in households during lockdown, the appeal of "good local products" increased during the lockdown period. With more time available, consumers favored raw, cookable products, with less concern for the convenience and speed usually sought in the daily routine of school, work and activities. Producers, restaurant owners and retailers were able to forge new links with consumers who were previously inaccessible, and to take advantage of the change in consumer habits and loyalty to products, brands and distribution channels. Forced to modify their supplies and diversify their consumption, which has often become exclusively domestic, consumers have had to reconsider the product and brand supplies with which they were previously familiar and find new ones[15]. From a social point of view, this refocusing on the local has allowed a rediscovery of nearby possibilities and has given increased visibility to previously marginal actors (AMAPs[16], producers' baskets, restaurant owners delivering meal trays to homes). This local life has also been enriched by an openness to neighbors, sometimes including mutual aid and conviviality, with those we used to pass by without noticing them and with whom moments of sharing such as aperitif-balconies or support for neighbors, isolated people or caregivers have developed. At the same time, the imposed social distancing and the use of digital

15. See: https://www.mckinsey.com/business-functions/marketing-and-sales/our-insights/a-global-view-of-how-consumer-behavior-is-changing-amid-covid-19.

16. AMAP: associations supporting small farming.

technology reduced the physical exchanges at the point of sale (with the producer, distributor, other customers and products). The use of farm drives increased during lockdown for health reasons and product availability. However, will this practice continue? Will local consumption also continue to grow? Local actors, municipalities and departments, who were very involved in the management of the health crisis, were momentarily a source of proposal for generating new tools and developing local networks. Metropolises have set up interactive maps to locate local producers and inform consumers[17]. However, at this stage, there is no indication that the Covid-19 crisis has led to a deep and ethically motivated recognition and attraction for these local sales outlets. How can virtuous behavior be sustained among these new users – initially attracted more by the motivations related to the context of the Covid-19 crisis and its consequences than by the virtues of quality and sustainability of local consumption – when the interests of health and the reliability of supply are less clear? Will the interpersonal relationships newly forged with these actors withstand the return of the imperatives of practicality, economy and available time, and the arguments of the large supermarkets and *low cost sector*?

7.6. What are the possible scenarios?

To conclude this epilogue, the main question concerns what will remain and what will emerge from this particular period of crisis. Fundamental changes were underway, such as the growing sensitivity of consumers to health, well-being, naturalness, the environmental impact of products, their origin and their production methods. New consumer practices or trends (increased role of digital technology, individualization of diets, return of homemade food, etc.) were identifiable before the crisis. Have they been extended, amplified or challenged? As mentioned in this chapter, the consequences of lockdown and consumer responses to a new context are not only very heterogeneous, but they also reveal numerous antagonisms.

In a September 2020 study[18], France Agrimer proposed different scenarios for "post-Covid" food consumption. These scenarios are distinguished by the different future modalities that the main variables related to the sanitary and economic situation, the management of the crisis by the authorities and the reactions of consumers could take. According to each of these scenarios, the consequences of the crisis on the evolution of food practices are very different. A return to a festive consumption that is starting up again in reaction to a difficult but ultimately

17. See: https://agriculture.gouv.fr/projets-alimentaires-territoriaux-les-initiatives-se-multiplient and https://www.chalontransition.org/annuaire-et-carte-interactive-des-producteurs-locaux/.

18. See: https://www.franceagrimer.fr/content/download/65171/document/Sc%C3%A9narios%20Consommation%20Post-COVID.pdf.

temporary crisis, or on the contrary, eating habits that are individualized and fragmented in a crisis that lasts and sees a succession of lockdowns. Economic constraints that encourage a retreat to local food and also, potentially, a recourse to cheaper food from abroad, depending on the recovery of international trade and the disposable income of households at the end of the crisis. The perception of food and conviviality as a health risk leading to isolation and the gradual disappearance of eating places, or a deep awareness of the need for a green transition towards greater responsibility and respect for health and ecology, in a context of persistent but relatively controlled crisis. The possibilities are multiple. After the health crisis from which it stems, the economic crisis that has begun and will continue will have consequences on food practices. The extent, duration and generality of this crisis will largely determine these consequences. The future of teleworking is also a determining factor in the future evolution of eating habits and conviviality. Between social distancing, the digital screen and the destructuring of time and professional/private spheres, teleworking also induces changes in eating habits, products consumed, rhythms, and of course the social dimension of mealtimes. The consequences on the products themselves are also very difficult to anticipate. We can also imagine a questioning of modernity and its acceleration, with trends such as slow food and the search for authenticity or the return of a form of nostalgia, as well as an increase in the need for practical, fast, functional or safe solutions, such as, at the extreme, an individual meal in a bar.

More broadly, the psychological and emotional consequences of the crisis are difficult to grasp. Will they lead individuals to seek more physical conviviality, particularly around meals? Will food, because of its characteristics as an "essential product", the need and the authorization to lower the mask, associating food with a certain form of freedom, crystallize a social link that would otherwise deteriorate in other aspects of daily life? The effects of this crisis and the changes in our societies that it has induced should occupy research on the consequences on our food practices for a long time. Reciprocally, the evolution of these food practices will probably constitute a vector of perpetuation of these sociological changes.

Conclusion

Guillaume LE BORGNE[1] **and Gilles SÉRÉ DE LANAUZE**[2]

[1] *Institut de Recherche en Gestion et en Économie (IREGE), University of Savoie Mont Blanc, Chambéry, France*

[2] *MRM Laboratory, University of Montpellier, France*

Can we still eat together? Can we still live our daily lives and celebrate together when our agendas, our goals, our values, our convictions and our health status diverge? Is it essential to eat the same food around a table or on either side of a screen? Under what conditions can we achieve this? Across seven chapters, and from several angles, this book has addressed the question of eating together: the benefits and conditions of implementation of physical commensality (Chapter 1) and virtual eating together (Chapters 2 and 6), the reactions of close relations to the adoption of a specific diet by an individual (Chapters 3 and 4), the influence these close relations have on the observance or non-observance of this diet, as well as the roles and contributions of communities of food practices and/or convictions (Chapters 2, 4, 5 and 6). Finally, the context of the current health crisis sheds light on these same questions (Chapter 7).

Eating together is not reduced to ingesting the same food around the same table: it takes many forms (eating together synchronously or asynchronously, physically or by being connected to the same virtual network, eating together by videoconference, eating or not eating the same dishes, in the same order or not, etc.). Moreover, eating or not eating together concerns all the stages – and therefore all the times and places – of food consumption (modes of food production, choice of products and planning of purchases, points of sale frequented, preparation of food, consumption, etc.). Thus,

Evolution of Social Ties Around New Food Practices,
coordinated by Gilles SÉRÉ DE LANAUZE.

this book also addressed "not eating together" (Chapters 1–5), whether or not to produce together (Chapters 5 and 6), whether or not to buy together (Chapters 4 and 5). In connection with eating together, these chapters provide an overview of new or established trends in food consumption. They are followed by an epilogue (Chapter 7) that puts the issues discussed in light of the exceptional Covid-19 crisis, and the changes that it has caused in food habits and concerns. This concluding chapter recalls the contributions of the different chapters of this book to the question of eating together and the associated consumption trends. It also proposes a complement to certain consumption trends and food innovations, and opens up new avenues of research, all the more necessary as the health crisis may herald a paradigm shift.

Eating together is desirable, and brings several benefits. Physical commensality brings structure to meals, leads to taking time for eating, and in fact leads to greater nutritional diversity and consumption of fresh products. Similarly, in a virtual setting, online communities participate (through the sharing of recipes, photos of dishes, tips and advice on nutrition, and also a list of all items consumed during the day) in the diversification of diet, which is a common point with physical eating together. In addition, eating together contributes to nutritional well-being through the proximity and conviviality associated with it. Eating together enriches the consumption experience and the social links that are developed, and generally contributes to an increase in the hedonic, symbolic and expressive values that are derived from it. In France, in a large proportion of cases, people usually eat together, in practice as well as in discourse and in the meanings attached to it, and even though food rhythms are in the process of being destructured. Common meals are perhaps one of the last places and times to spend time with our loved ones, on a daily basis or on the occasion of festivities. It is also, probably, one of the last moments to be held sacred, especially with regard to the use of cell phones.

Eating together can nevertheless be a source of stress and conflicting norms and can generate a set of social and identity constraints. For some people, eating alone may be a way to escape from overly burdensome social norms, to resolve incompatible beliefs or to respect health imperatives. The difficulties of eating together may thus result from a latent conflict between a functional physiological approach to food and a holistic approach that integrates its social and symbolic dimensions. Difficulties may also result from the conflict between a binary and categorical vision of food (i.e. there are intrinsically good products and intrinsically bad products) and, on the other hand, a contextualized, relative and differentiated vision of it.

C.1. Producing, buying, eating together on different scales

The question of eating together also concerns the other stages of eating, beyond the single act of ingestion. It integrates "producing together" and buying together, involving different scales (individual, family, friends, colleagues, territory, etc.) where it also results from collective and individual choices. Through all of their food practices, individuals shape their identity, situating and asserting themselves in relation to others. One can eat with one's close circle (family, friends), with one's colleagues, next to strangers, on different occasions (daily, festive, periods of crisis or remission for a chronic disease, etc.) and with different individual and collective meanings (eating for utility (for hunger, to take care of oneself), for ritual, for the pleasure of discussing what one eats, etc.). All this represents a great diversity of contexts and situations for eating together. Eating together can also be considered at the scale of a territory, which individuals (consumers, producers, distributors, etc.) share and occupy together. The terroir refers both to a set of products and to a human community sharing an identity around these products. Eating local is a way to support local producers: here, eating together is closely linked to living together in the same territory, to stay close to products consumed and their producers, with the desire to avoid the relocation of production. The diversity of local products is a mirror of national cohesion. Eating locally is "a way to reweave the links between all the stakeholders of the territory, especially between producers, distributors and citizen-consumers". Consuming local products in a given area can be done in order to give an identity to this territory and its occupants, and to create a community. Collective sales outlets such as producers' stores are places where people sell together. In other cases, it is possible to "produce together" and differently, as the example of the halal meat market shows. According to the French Ministry of Agriculture, 62% of slaughterhouses practice ritual slaughter at least occasionally. In some cases, ritual slaughter tends to be closer to the "standard" of conventional slaughter. In other cases, it is the opposite, and there is hyper-differentiation. The consumption of halal meat is a differentiating practice. Halal meat is mainly sold through traditional channels that are entirely halal (consumers of halal products and those of non-halal products do not buy together). However, it is also sold in supermarkets (in this case, halal and non-halal consumers make their purchases in the same place).

Some consumers of halal products avoid supermarkets, such as one consumer in this example who avoids supermarkets because halal meats are next to non-halal meats, even pork. Other consumers of halal products feel uncomfortable with the gaze of some of the customers. On the contrary, many consumers buy halal meat in supermarkets, as recent trends and figures confirm. We can thus see that a specific market can give rise to a variety of processing and distribution channels, and to very distinct consumer reactions to "buying together".

The question of "buying together" also applies within diet communities: members ask other members to describe their eating habits (food, preparation), and also their buying habits: Where do you buy your products? To eat like someone else is also to buy like them (and thus, often, in the same type of outlet, to find all the products in question).

C.2. Eating together: between identity construction and social influences

In fact, eating together covers a very wide variety of practices and motivations: eating together as a way of getting together, eating together with the esthetics of the meal as one of the central subjects, eating together because it is necessary to eat and because we are together and so on. In short, it can be a means, a necessity, an end, or it can be incidental. It requires material elements, a given space and time, knowledge, skills (culinary and social), and finally meanings that, when not shared, can complicate this practice or this moment. Motivations and objectives, convictions and medical needs may be multiple and not always shared.

Criticism of meat consumption (pollution, significant use of resources, animal suffering) makes many people uncomfortable with the contradiction between habits, positive meanings, and a taste for meat, and these known harms. For some consumers, the resolution of this contradiction is to become vegetarian. For others, there is relativization, decredibilization or trivialization as avoidance mechanisms in order not to stray from the omnivore diet. Finally, there are "in-betweens", such as flexitarianism, unimaginable for some and the only possible compromise – at least initially – for others. In addition, tensions can emerge between groups, for example between vegetarians and the rest of society. The opposition can be between the vegetarian community and the society, and it can be manifested on a different scale: between an individual and their environment. Vegetarianism is still marginal, and this practice is out of step with the dominant normative model. The vegetarian must then manage this "normative dissonance" in a context of vegetarian/non-vegetarian opposition. A typology is proposed which distinguishes three categories of vegetarians according to the normative dissonance: the integrated (weak dissonance) possibly by flexibility of the diet, the isolated (avoidance strategies to survive) and the opposed (assert themselves, and try to convince).

Consideration of the future consequences of a food choice can sometimes clash, at the time of the meal, with the motivations of certain diners who are more interested in the immediate consequences of the meal (search for pleasure, conviviality, limited time to eat, etc.). Thus, the specific diet that is vegetarianism can be based on health motivations (generally internal motivation), animal welfare (external), ecology (external) or religious (internal). Sometimes, the rejection of a food category can go beyond the cognitive-emotional, and be visceral, as shown by the example of an interviewee who forces herself to eat mollusks with her family, and that her body (as well as her mind) finally rejects.

Not eating alone but with others can give rise to conflicts of norms, and dissonance between attitudes and actions. These conflicts and dissonances, often latent, are revealed through eating together, which can then be perceived as a problem.

The gaze of others, and normative pressure, are influential elements in food choices. For individuals with "specific" dietary habits, this can be particularly sensitive, both from the point of view of descriptive norms (what others do and what I can therefore consider as "normal") and injunctive norms (what I should do, what I am told to do). In the previous versions of the PNNS (*Plan National Nutrition Santé*, French National Nutrition and Health Plan), the (prescriptive) dietary norms may have led to the stigmatization of individuals or groups (social, cultural, etc.) by not taking enough account of the particularities of these groups. As the majority of behaviors are not particularly healthy, with median intake of sugars, processed fats, and ultra-processed products well above WHO recommendations, there is often a conflict between the dietary norms of the person implementing a special diet – a low-calorie diet or exclusion diet – and the norms of the majority diet. Dieters may often feel "abnormal" and indeed be seen as such by others. However, the situation can be reversed: it is the "classic" diet that is "abnormal", criticized, and having to justify itself (e.g. conflicts between vegetarians or vegans and carnivores).

The reactions of the people around them to specific diets (whether low-calorie, gluten-free, vegetarian, raw food, etc.) can be very varied, ranging from hostility (lack of understanding of the motivations, feeling that one's own practices and convictions are being challenged, or at least questioned), to positive reactions and a desire to follow the example, through criticism (concern for the health of the person who excludes certain foods) or more or less neutral tolerance. In the case of the exemplary nature of the specific diet, eating together makes it possible to trigger a change in the practices of other members of the group. It is therefore not always the individual on the special diet who abandons their diet in order to conform to the group norm: the opposite can happen.

Depending on the type of diet and the underlying motivations, the latter will be more or less understood and more or less accepted by those at the table. In the case of a chronic disease, in periods of crisis a person will implement strict adherence to the diet, and in periods of remission they will be able to eat some foods that are forbidden in periods of crisis. These variations can be difficult for other people to understand: if you allow yourself to deviate once, why not the next time? In the case of a strict vegetarian following this diet without exceptions, there are no variations and the rule does not fluctuate, nevertheless the problem may lie elsewhere: when the chronically ill person follows this diet for their health, this vegetarian may, for example, follow this diet for ethical reasons, thus confronting the omnivores with their responsibilities as eaters, which may generate discomfort of another kind, as explained above.

Beyond the advantages of commensality presented in the introduction, and beyond hedonic well-being, would the well-being resulting from eating together come from the pleasure of celebrating practices with common meanings (eudemonic well-being), and from the success in the ability to circumvent constraints in order to "manage to eat together"? The concessions can go both ways. Indeed, a person on a diet deviates in order to adapt to what the table consumes, or a person who does not impose restrictions on themselves makes an effort to adapt the meal in order to cater to the restrictions of another, or to consume the products proposed by the latter and which may be outside their "classic" habits. But sometimes, the complete and permanent change of diet can only take place once the initial reference group has been left. Finally, if one is unable to maintain the practice (e.g. because of material or social difficulties), the last choice available may be to adapt one's convictions or beliefs to one's practices in order to reduce the state of dissonance. In the case of vegetarianism, this may be to consider certain modes of production or certain types of meat as ultimately acceptable. In the case of low-calorie diets, it may be to tolerate deviations from the set standard and to consider that their impact on the goal is marginal.

The consumer trends presented in this book are the result of emerging concerns and convictions, in line with health, social, environmental, ethical and moral issues.

A restricted diet can be either chosen (values, convictions, health objective not strictly necessary for medical reasons, wish to accompany a loved one by sharing their diet, etc.), or imposed (chronic illness, allergy, etc.). In both cases, it is a question of an in-depth modification of eating habits and behaviors, up to the organization of daily life and certain non-food choices. For example, deciding not to eat meat goes beyond a simple change of diet: it modifies habits and social relations linked to food. The vegetarian diet has an impact on the organization of purchases, the preparation and eating of meals, both at home and away from home, with the material and social organization that must be put in place. Solutions must be found to many material, organizational and relational questions. New products must be found, as well as new places of supply, new recipes and so on.

Social influence then prompts the person to adapt their behavior, their attitudes or their beliefs to their social environment. Food consumption is subject to this social influence, both informational and normative. The reference group is a source of information, norms, values and aspirations that validate opinions, attitudes and behaviors (value-expression influence). In online communities, "the values of the group and the individual are identical" (which is not a given for the initial reference group). The influence is reciprocal, consented to, and even sought after.

The social environment and the social support it provides play a key role in nutritional compliance (or non-compliance). The close circle does not necessarily share the constraints of the individual who is following a specific diet, but can help to respect them, or, on the contrary, encourage non-compliance. The people around them may show a lack of understanding. Adaptation and support can strengthen bonds. Their absence can cause them to break down. Becoming vegetarian, becoming vegan, is a long, complex and difficult process: new knowledge to be acquired, modes of supply to be modified, new habits to be adopted, renunciations, reactions and influences from close and less close people. This choice is based on numerous motivations and underlying convictions that are personal. The practices implemented to follow this diet – as with other diets – are part of a life course with its constraints, its context, social contacts and an evolving course subject to hazards. The process of adopting and maintaining the vegetarian diet is strongly influenced by the relational aspects of meat consumption, in addition to material and psychological constraints.

However, the close circle of friends or the initial reference group does not necessarily provide a favorable environment for following a specific diet trouble-free. The community (weight-loss, allergy, food intolerance, chronic illness, vegetarian, religious, etc.) plays a very important role in resolving dissonance, in following the diet easily, in providing the necessary support and reassurance and in proposing an identity model.

In this book, different communities defined by food practice choices, based for some on strong convictions, are discussed. A community (of identity, of practices, of convictions, etc.) is an abstract or concrete entity, physical or virtual, spatially extended or more compact, more or less formalized, that gathers individuals. With the Internet, the development of online communities, as "gatherings of individuals who have common interests and goals" has appeared. These communities are a place for sharing: recipes, photos of dishes, lists of all the ingredients consumed during the day, tips and advice on nutrition, personal states of mind and so on, and are thus, more broadly, a place for living. Health communities provide informational (opinions, advice, tips) and emotional (encouragement, mutual comfort, empathy) social support, motivation, information and sharing of experiences. Usually, asynchronous conversational exchanges take place (contrary to the physical case), around the same focus: weight loss.

Whatever the purpose of the community, individuals are free to join or not, unlike the physical reference group – usually the family – which is not chosen. Participation in the physical and virtual communities presented in this book can be based on a personal goal to be reached (a healthy weight) and a state to be maintained (with a diet that may be different from the transitional diet towards the goal). It can also be based on dietary practices that are stable over time (gluten-free diet, vegan consumption, etc.). Other food trends not addressed in this book involve complex processes of transformation of consumption patterns, such as the "practice" of zero waste. The success of these practices is very often based on an identifiable and active community. It is indeed very difficult, without a community to rely on, to maintain the discipline and the rigor that allow the observance of the diets presented in this book. And the complex motivational and behavioral process of becoming vegetarian, for example, cannot be easily understood by those who are not.

Communities where food is the focus of discussion are groups whose members largely have a high level of dietary involvement, in contrast to the heterogeneity found in society at large. They can be a way to "eat with" people who share our values and/or norms and/or goals, when this is not the case with our tribe. We seek support (and advice) from people who have the same goal, who understand us and who we know will not judge us. They can also be a help to still "manage to eat together with our loved ones", as shown in the example in Chapter 2 of a woman asking for advice from other community members in order to eat out with her spouse while limiting deviation from her diet. There is reciprocity of support, because supporting others, motivating them, motivates oneself. Members are interdependent, with both a need to be supported and a need to support others.

Online communities can allow individuals to eat together online, and in this way to eat together better than with their physical groups. But through external social support, they also allow people to eat more easily with one's relatives: the support and reassurance of a chosen virtual community then improves the conditions of eating together physically in the non-chosen home. Different roles of the community are identified: functional, hedonic, social and political/moral (specific to vegetarian compared to the other communities in this book). Within the community, one finds information on the issues, on the practices, assistance, advice, examples, in order to reduce the risks when making choices. We develop our interest, our curiosity, we make discoveries in a friendly atmosphere. Communities of conviction (e.g. the vegan community) can lead to debates and action to convince the rest of society. Communities influence the behavior of their members, and also that of non-members.

But being part of a community also provides identity reinforcement, through belonging to a defined group, social recognition and defense against external aggression. It is a place of awareness, reinforcement of beliefs and supporting discussions with non-vegetarians. Resorting to the community influences the way in which the vegetarian will live their practice within society.

Following a particular diet can be seen as an "anomaly" and the person following it can be perceived as sick, or marginal. When we eat with colleagues, friends or family, our diet says something about us that we do not necessarily want others to see. This is especially true when it concerns health, whereas for vegetarianism we may wish to be seen by others (e.g. to provoke debate) or on the contrary we may not wish to (e.g. because we do not wish to mark our difference or to debate with such and such a person or group).

The observation of these differences from others, added to convictions and/or constraints that are often strong, leads to the need to be recognized. For this reason, vegetarianism becomes an identity (one is not vegetarian by chance, one is not vegetarian easily, one is theoretically not half-vegetarian, it distinguishes us from the majority and defines us in opposition to this majority). From then on, there is a strong social dimension. The individual defines themselves as a member of the community on the basis of similarities with other members and corresponding differences with the outside group (non-members). The perception of oneself as a vegetarian can differ greatly from person to person. Some people do not feel part of a community, nor do they want to. Others feel part of it, and need it. It depends on the desire to identify with the group, and the normative relationships with the rest of society. There is identification with a model and identity desirability of this model for some, when others do not seek this identity, do not need it or even judge it negatively.

Similarly, Chapter 5 reminds us that halal meat purchases are very much marked by religious events, and that it is a differentiating religious practice. The main determinants of the act of purchase are the religiousness of the act, the effect of community membership and the halal label. We can thus see that in the case of a community of religious practices, community membership also has a significant influence on behavior.

Finally, eating together in online communities is, by definition, based on the use of digital media. The Web makes it possible to eat together at a distance, and not only in communities of practice. Appetizers or meals with family or friends via videoconference already existed before the spring 2020 lockdowns, but these practices became common as a result of them. Digital communication terminals (phones, tablets, etc.) present paradoxes and ambivalences. On the one hand, they allow a connection to close relations (including "eating alone together", though probably a degraded experience), to communities (notably of practices) as we have seen previously. On the other hand, they parasitize interpersonal relations in particular at the table, degrading the experience of physical commensality and what can be drawn from it. Thus, digital technology allows us to be together when we are alone (the experience of the lockdowns linked to Covid-19 has shown this), but can conversely isolate us when we are together (at the table or during the preparation of the meal or at any other time).

The questions related to the use of digital technology also concern local supply networks, in connection with or in opposition to the network that is the Web. Does the Web make it easier to access local products? It is difficult to answer this question at the moment: digital networks certainly allow the grouping of a hitherto fragmented offer of local products, but they modify – and can degrade – the experience of consuming these local products, if only by reducing the opportunities for physical contact with producers or with the place of production. Moreover, these issues are not the same in urban and rural areas, which are nevertheless interdependent, and whose relations are part of an era in which there is a general desire to "reconnect metropolises and their territories".

Finally, and more generally, it would be interesting to explore in future research the commonalities and differences between physical eating together and remote eating together. The activities performed in parallel to food ingestion are not the same in physical eating together (oral discussion, touching and seeing, exchanging dishes, managing the table) and digital eating together (reading and writing messages, videoconferencing, consulting web pages, or any other activity that is not necessarily digital, etc.). Aspects related to rhythms may also differ (synchronism, meals generally eaten more slowly than alone, vs. asynchronism and perhaps meals eaten more quickly?), as may the physiological consequences and the different symbolism attached to these practices. From an experiential as well as a physiological point of view, the comparative and systematic study of ways of eating together physically and virtually – discussed in this book – is a promising avenue of research.

C.3. Eating together: adapting practices and offers

It is entirely possible to eat together without eating the same thing. The preferences and tastes of the guests can be distinct, and the simultaneous presence on the table of a variety of dishes that the guests can consume (or not) can promote conviviality despite the differences in choice and diet. The offer and its evolution then play a major role. The inner circle and the community bring understanding, adaptation and social support to the person not following the majority diet, and the offer can play a facilitating role.

Food innovations are taking place at the product level (vegetarian barbecue sausages, halal "wine", vegan and gluten-free powdered meals, etc.), both for those who follow a special diet and for those who do not. Indeed, many players are adapting to the evolution of diets, by proposing an offer that facilitates their adoption and monitoring (product and service innovation), and by anticipating favorable developments in society (towards veganism, organic, etc.). The food industry offers dietary products (gluten-free pasta, vegetarian lasagna, etc.) without necessarily adhering fully to the corresponding "community". Several specific diets discussed in this book are exclusion diets: they involve not consuming one or more categories of products. The fact that this is "without" rather than "with" (e.g. consumption of dietary supplements, such as protein), is not anecdotal. Going without a product (or having someone refuse it) is not the same as consuming an additional product (or having someone consume an additional product that is not part of our diet). From a psychological and social point of view, this distinction has consequences that could be studied. In addition, most of the innovative products in the gluten-free, vegetarian, halal, etc. segment are ultra-processed products, which opens up avenues of research in terms of perceived naturalness and health risk, and thus acceptance of these products by people following a specific diet.

The French brand, Feed., which sells food in bars and powder (to be mixed with water), is an example of product innovation that could be of interest to people with specific diets (vegan, gluten-free, dairy-free, low-calorie diet, etc.) as well as omnivores (when traveling, quick meal at work, etc.). A bar or a dose of powder is, according to the brand, "equivalent" to a meal, because it contains the recommended quantities of macronutrients. This is a functional and reductionist way of looking at food, that is, by reducing food to the sum of its parts, and a meal to a set of nutrients whose quantities would be balanced whatever the form (raw product, mixture of extracts of unstructured products, etc.) and the source. These easily transportable and preservable (because they are dehydrated) meals can be a way to eat alone and quickly at work or on the move, as confirmed by the brand's communication and consumer reviews. However, it is also a potential emergency solution for people following a specific diet. Having one of these "meals" with you in case of need ensures that you can eat something controlled and that can help you occasionally. For example, these could be useful when taking a trip when you do not know the availability of suitable food in advance (train, plane, etc.), or if you have to eat with other people around a meal that does not respect certain constraints. This inclusive argument is put forward by the brand when presenting its product. This type of innovation, somewhat similar to what is observed in the development of digital practices, is not without certain contradictions in the development of "eating together". On the one hand, they allow a person following a specific diet to easily adapt to an environment that is not always adapted to this diet, but, on the other hand, they can lead more and more eaters to be satisfied with a powdered meal eaten alone and quickly, and thus contribute to further reducing meal times and conviviality. The products will then have to solve these intrinsic (ultra-processed, poor taste and texture, "it's not a food", etc.) and extrinsic (degraded consumption experience, "for me eating is not that", etc.) obstacles, in particular by providing the "naturalness" that may be lacking in these highly processed products. Finally, eating together in the case of distinct diets does not only rely on an increase in the number of dishes at the table or on product innovations, but also on social and service innovations. Many restaurants, for example, have adapted by offering a number of vegetarian dishes in an explicit and easily identifiable way. Public canteens are also increasingly offering a separate vegetarian dish, or the dish of the day without meat for a lower price. Nevertheless, these approaches remain scattered and, if they are spreading for vegetarianism, what about other practices (gluten free, halal)?

Diets have also led to trends in distribution. The case of halal allows us to look at complementary aspects, linked to the place of purchase and product range. While the traditional channels (and in the first place, the large-scale distribution) propose new products that facilitate "eating with others" and "eating like others" (foie gras, organic products, ready-made meals, wine, etc.), the halal segment includes a demand of difference. People consuming halal products can eat with others and eat "like the majority", through purchases of common brands in common outlets (halal in supermarkets, in fast food restaurants (Quick), halal range of Fleury Michon), and/or eat differently through specific practices and purchases along traditional channels and/or halal brands. Buying and eating halal is a specific act of consumption, which has not prevented the development of a standardized offer in mass-distribution, and the traditional channel itself has become industrialized/financialized. Nevertheless, some consumers resist this standardization, making room for a hyper-differentiated "authentic halal". As in the case of halal, people following exclusion diets (gluten-free, lactose-free, vegan, etc.) have the choice between undifferentiated universes such as that of supermarkets – with a standardized offer located, or not, on specific shelves (gluten-free shelves, organic shelves, sub-shelf of vegetarian meat substitutes, etc.) – or specialized stores (organic grocery stores, producer stores, vegan stores, gluten-free bakeries, zero waste bulk stores, etc.). The place of purchase and supply overlaps the practices in defining segmented, even community food consumption perimeters, leading buyer-eaters towards an increasingly specialized offer, or forcing them to multiply supply points. Do specialized stores and specialized departments in supermarkets represent a form of spatial segregation of consumer segments, reinforcing the spatial segregation that is already taking place through living areas? Does not buying together or eating together contribute more broadly to no longer living together (between different social classes, different ethical or religious convictions, etc.)?

Distribution is also influenced by the arrival of mobile food applications such as Yuka (the most widespread), The Gluten Free Scanner (gluten-free diets) and Veg'Check (vegetarian diets). These applications – based on the scanning of product barcodes with a smartphone – accompany consumers in their purchases, facilitating their decision-making (displaying sorted and condensed information, with an overall score assigned to products and/or customizable filters to exclude certain foods). Their users – who in the case of generalist apps like Yuka are not necessarily following a particular diet – generally trust these apps more than they might trust the brand or the retailer. The use of these applications is first of all informative and individual. But it can also be social, on the one hand, during exchanges between friends and family around the evaluations displayed by the app and the food choices to be made, and on the other hand, because some nutritional apps integrate a collaborative (such as Open Food Facts) and thus community function. Several brands – including retailers – have already reacted by modifying the formulation of their products so that their rating exceeds the 50/100 threshold on the Yuka app, for example. But the impact of these applications goes beyond that: whether this use is individual or social, it modifies the purchasing process of users, and leads retailers to question their product range, merchandising and customer journey strategies.

Finally, the distribution and consumption of local products are giving rise to multiple innovations. Until the middle of the 20th century, consuming locally produced food was the predominant form of consumption, before the generalization of international trade in export crops and manufactured food products, and long and ultra-long distribution channels. "Return to the local" could therefore give the impression of a return to old practices, as opposed to an innovation. On the contrary, the territorial product is now based on innovations in production and quality processes, as well as in distribution. The proximity of actors – threatened by demographic dynamics and changing lifestyles – is organized, on scales and with constraints that are different from what could be observed in the old forms of "consuming locally". The intense work of networks of actors has led to the emergence and support of numerous short circuits (markets, producers' stores, AMAPS) with unique forms, in the sense that they are adapted to the geographical, social and economic realities, and to the expectations of the different actors (local producers and consumers at different levels of commitment). Innovation is therefore not only in the human and technical processes of production (appropriation of know-how, traceability and quality, etc.) and distribution (logistical aspects, online platforms, etc.), but also in the social innovations that this "consuming local" calls for or even requires.

C.4. Eating together: What changes are expected?

In this book, several aspects of eating together have been addressed: the description of practices and situations, the relational impacts, the normative dissonance induced by contradictory social influences between different groups, the adaptation of the offer to better fit concerns and particularities linked to different convictions, or the variable perimeters of the contexts of food consumption and production. New trends are changing the traditional framework of commensality, its conditions and meaning, whether at home, at work or in restaurants. The time pressure and constraints of working life and increasing nomadism threaten "eating together", but they are counterbalanced by new opportunities, new offers and new services. Some of these opportunities are brought by digital technologies and the information sharing and exchanges they allow. Even though sometimes seen as vectors of isolation and time-consumption, they contribute to the reinforcement of convictions around food practices and in the resolution of concerns. The exchanges that they allow are developing more and more within election groups, and around convictions that are strongly relayed via social media, expression spaces for these communities that play a support role to their members.

Thus, the book contributes to a better identification of social practices around food, useful for market strategies of actors and public policies aiming to improve health and food well-being, of which conviviality appears as an important potential antecedent. It also opens up new avenues of research, in relation to societal changes – both fundamental and linked to the health crisis – and the concerns and behaviors that arise from them. In particular, the book addresses conviviality in the era of digital tools. The latter allow new ways of sharing information and moments of conviviality around meals, while at the same time enabling individuals to become isolated, as screens reduce moments of physical sharing in time and space. When it comes to food, it is possible to be connected to both the Web and one's territory, and digital means allow the extension of physical links, improving the identification, valorization and access to local products or specific diets. But further research is needed to deepen the study of the factors conditioning the benefits of digital possibilities, especially in terms of motivation and affordability. At the same time, the resistance of consumers to techniques that could counteract the proximity gained with the local fabric is also to be studied. The point of view of producers, who may see it as easier and more direct access to a local clientele, but also and conversely as a screen to a real relational approach, not to mention the practical difficulties and the perceived costs, should also be investigated.

Some themes around the social links of food have not been addressed in this book, and deserve greater interest from researchers. In particular, dependency and solidarity in an ageing society are one of the major angles of food consumption study. Numerous collaborative initiatives, sometimes linked to direct distribution channels, have been created or strengthened as a result of Covid-19, in order to deliver food shopping or prepared meals to vulnerable people. For the latter, it is as much a question of nutritionally indispensable supplies (in a context where the vulnerable must limit their trips to the store because of the risk of contamination) as it is of social links and conviviality, which are just as vital. In this double context (long-term aging trend and a pandemic of unknown duration), social food links take on a reinforced meaning and importance. In addition, out-of-home catering (restaurants, canteens, etc.) has not been specifically addressed in this book. The values, norms, social influences and identity affirmation in this context are different from consumption at home, and the question of eating together is posed in a complementary manner.

Several of these chapters have highlighted the growing importance of groups and communities, to which membership helps, shapes and sometimes conditions the evolution of food practices. As a source of information and shared pleasure, as well as a support for underlying convictions, these groups complement the family or traditional circles of commensality, to propose other forms of "eating together". It also appears that these communities, often with very formal outlines based on moral, religious or ethical convictions, rely on a product offer specifically designed for them. This can be done in an extremely rigorous and regulated way, as in the case of halal meat, but it can also be observed in a sanitary approach with gluten-free products or in an ethical or philosophical approach as in the case of veganism. In all these examples, the question of quality certification arises, and also that of the link between the credibility of the offer, compliance with the diet and, in connection with the relationship to the community itself, the question of the normative pressure generated by the community.

The study of these communities of adoption, of the phenomena of social identification and of the cross-influences that they give rise to, should also be continued. If food is part of identity and belonging, the analysis of intra-group, extra-group and inter-group relations could help in understanding food behaviors. In the case of vegetarianism, we understand how the relations between vegetarians and non-vegetarians can affect the process of adoption and maintenance of the practice. The relationship of vegetarians to the community as an identity component and their perception of a more or less strong opposition between the vegetarian approach and the traditional habits and vision of society concerning meat can also affect the process of adoption and maintenance of the practice. Such mechanisms are also suggested in the case of health and weight loss diets. These questions are important for the actors of these markets. They make it possible to envisage new avenues of segmentation and innovation, in terms of products and also of communication and promotion of these specific offers. Can companies claim to be expanding their outlets through a gradual diffusion within society – in particular through forms of standardization of an initially singular offer – and through support for the corresponding communities? On the contrary, can they innovate within the framework of a strategy of hyper-segmentation by crossing the motivational contours of several groups? For example, can we believe in the development of organic halal? Or gluten-free vegetarian products or products strongly anchored in a local territory?

Finally, will the increasingly visible multiplication of specific diets – crossed with varied food styles – lead eaters to forget the idea of a single dominant standard, and to accept the complex coexistence of multiple norms and practices? In the absence of a standard for all, this book has shown that it is possible – and difficult – to eat together, and that it can be achieved mostly with repeated – individual and collective – adjustments to evolving norms and needs.

List of Authors

Laurie BALBO
Grenoble École de Management
France

René Pierre BEYLIER
MRM Laboratory
University of Montpellier
France

Foued CHERIET
UMR MoISA
Institut Agro-Montpellier SupAgro
France

Margot DYEN
Institut de Recherche en Gestion et en Économie (IREGE)
University of Savoie Mont Blanc
Chambéry
France

Fatiha FORT
UMR MoISA
Institut Agro-Montpellier SupAgro
France

Steffie GALLIN
Montpellier Business School
MRM Laboratory
University of Montpellier
France

Andréa GOURMELEN
MRM Laboratory
University of Montpellier
France

Félix JOURDAN
UMR Innovation
INRAE
Montpellier
France

Guillaume LE BORGNE
Institut de Recherche en Gestion et en Économie (IREGE)
University of Savoie Mont Blanc
Chambéry
France

Marie-Christine LICHTLÉ
MRM Laboratory
University of Montpellier
France

M'hamed MERDJI
Montpellier Business School
France

Thibaut NGUYEN
Trends & Prospective Department
Ipsos
Paris
France

Andry RAMAROSON
CHROME Laboratory
Centre Universitaire de Mayotte
and
MRM Laboratory
University of Montpellier
France

Gilles SÉRÉ DE LANAUZE
MRM Laboratory
University of Montpellier
France

Lucie SIRIEIX
UMR MoISA
Institut Agro-Montpellier SupAgro
France

Erick SUAREZ DOMINGUEZ
MRM Laboratory
University of Montpellier
France

Index

V, W